*A Manual
of Anatomy and Physiology*

A Manual
of Anatomy and Physiology

Laboratory Animal: THE FETAL PIG

ANNE B. DONNERSBERGER
Moraine Valley Community College

ANNE E. LESAK
Moraine Valley Community College

MICHAEL J. TIMMONS
Moraine Valley Community College

D.C. HEATH AND COMPANY
Lexington, Massachusetts   Toronto

Published simultaneously in Canada.

Printed in the United States of America.

International Standard Book Number: 0–669–01490–7

Preface

Greater appreciation of the human body can be attained by learning the structural and functional units that comprise the body. Study of anatomy and physiology enables one to see systems serving the whole, and the whole serving its systems, in an extraordinarily integrated manner. The relationship between anatomical structure and physiological functioning, however, can be confusing to the beginning student. This laboratory manual facilitates the identification and understanding of anatomical structures and the basic functional organization of the human system.

A Manual of Anatomy and Physiology (Fetal Pig Edition) is designed for the beginning student in a health science or related program. It presents primary and practical information relevant to the following health science occupational areas: medical technology, nursing, respiratory therapy, medical record-keeping, operating room technology, dental hygiene, pharmacology, biomedical technology, physical and speech therapy, physician assistance, and special education areas. Medical, legislative, and social advances have precipitated the need for adequately trained personnel in medical and medically related fields. It is for students in these areas that this manual has been designed, since most anatomy and physiology laboratory manuals provide little, if any, information needed by allied health students.

A number of unique presentations have been designed to facilitate learning. The illustrations provide visual impact for the textual statements. Three-dimensional illustrations and black-and-white photographs are placed in proper perspective to furnish visual reinforcement. Especially unique are the medical terminology exercises; the tables describing human bone markings with corresponding three-dimensional line drawings; the tables describing fetal pig musculature and its correspondence to human musculature, with related illustrations; and photographs of the internal organs of the fetal pig. Throughout the manual, terms that appear in **boldface** type are defined in the glossary-index at the back of the book.

The manual includes fourteen integrated units on the anatomy and physiology of the major organ systems. Included in the format of each unit are the following sections: purpose, objectives, materials, procedures, and a discussion (set of questions) related to the unit content. Within each procedure section, the student will identify major organs, study the gross anatomy of selected organs, examine the histological features of selected tissue, and perform physiological exercises related to the organs or systems of study.

The authors have designed this manual to illustrate, through carefully chosen exercises and correlated photographs and illustrations, specific anatomical and physiological principles especially relevant to students in health sciences programs. In order to further stimulate student involvement, selected photographs are partially labeled, allowing the student to complete the identification of unlabeled structures. These unlabeled structures are correctly identified in the Illustration Appendix on page 285.

The authors intend this manual to provide students with a basic understanding of anatomy and physiology as well as the necessary background for further study and work in the health care delivery system. In view of the authors' desire to improve *A Manual of Anatomy and Physiology (Fetal Pig Edition)* in response to teacher and student needs, they invite constructive comments from those using it.

A companion volume of *A Manual of Anatomy and Physiology* that features the cat as the anatomical subject is also available.

Acknowledgments

We wish to express our appreciation to the following individuals for their help and encouragement in the preparation of this laboratory manual: to our students, who demonstrated the need for a manual of this design; to Calvin C. Kuehner, Ph.D., for his contribution to the laboratory exercise on osmosis; to Mary E. Bannon for her suggestions regarding laboratory design; to Pat Oakes, who drew the three-dimensional illustrations; and to Peggy Svitanek for typing the manuscript.

Photomicrographs were taken with a Carl Zeiss Research Microscope made available through the courtesy of James Sullivan at Eberhardt Instrument Company. We thank Chuck and Rick Simak of Rayling, Inc. for hand processing each photograph.

We warmly acknowledge the editorial and production assistance of D.C. Heath and Company. Finally, a special note of thanks to our families, whose patience, encouragement, and understanding contributed significantly to the completion of this manual.

Anne B. Donnersberger
Anne E. Lesak
Michael J. Timmons

Palos Hills, Illinois

Contents

Illustrations

*Drawing
†Micrograph or Photograph

UNIT I *Medical Terminology*

PURPOSE

The purpose of Unit I is to acquaint the student with medical terminology with respect to body structure and medical procedures.

OBJECTIVES

In order to complete Unit I, the student must be able to do the following:

1. Demonstrate proficiency in using terms describing body directions, planes, and abdominal regions.
2. Utilize prefixes indicating location, direction, tendency, and number.
3. Use prefixes that identify organs.
4. Use suffixes correctly that are commonly associated with various medical areas.
5. Construct medical terms from common descriptions.

PROCEDURE

EXERCISE 1

Terms Describing Human Body Directions

Anatomical position refers to the human body in a standing position, with the palms facing forward. These terms are used in describing human body directions in anatomical position:

1. **Superior** (or **cranial**): Toward the head end of the body. *Example*—The shoulder is superior to the hip.
2. **Inferior** (or **caudal**): Away from the head. *Example*—The knee is inferior to the elbow.
3. **Anterior** (or **ventral**): Front. *Example*—The nose is anterior to the back of the head.
4. **Posterior** (or **dorsal**): Back. *Example*—The shoulder blades are on the posterior side of the body.
5. **Medial** (or **mesial**): Toward the midline of the body. *Example*—The great toe is medial to the little toe.
6. **Lateral**: Away from the midline of the body. *Example*—The little toe is lateral to the great toe.
7. **Proximal**: Toward or nearest the trunk or point of origin of a part. *Example*—The elbow is proximal to the wrist.
8. **Distal**: Away from or farthest from the trunk or point of origin of a part. *Example*—The foot is distal to the knee.

In a quadruped organism, such as the pig, the cranial portion of the body is referred to as *anterior* and the caudal portion *posterior*. Likewise, in a standing position, the quadruped dorsal or upper surface is *superior,* and the ventral, or belly surface, is *inferior.*

1

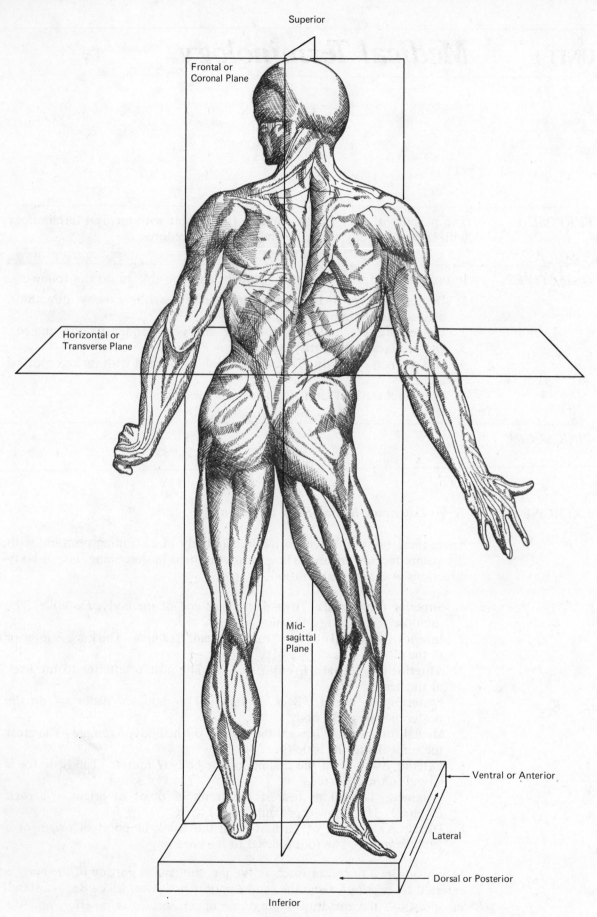

FIGURE 1 *Planes of the body*

EXERCISE 2 *Terms Designating Planes of the Body*

These terms designate planes of the body (see Figures 1 and 2):

1. **Sagittal**: A lengthwise plane, running from front to back, that divides the body or any of its parts into right and left sides (not necessarily equal). **Median Sagittal**: A plane that divides the body or its parts into right and left halves.

2. **Coronal** (or **frontal**): A lengthwise plane, running from side to side, that divides the body or any of its parts into anterior and posterior portions, or, in the quadruped, into superior and inferior portions.

3. **Transverse** (or **horizontal**): A crosswise plane that divides the body or any of its parts into superior and inferior parts, or, in the quadruped, into anterior and posterior portions.

FIGURE 2 *Fetal pig, coronal, serial sections, showing internal organ detail in situ*

1. Facial Coronal Section, Showing Brain, Optic Orbits, Tongue, Mandible

2. Shoulder Section, Showing Cerebellum, Spinal Cord, Vertebrae, and Trachea

3. Thoracic Section, Showing Lung Lobes and Heart

4. Abdominal Section, Showing Liver Lobes and Stomach

5. Lumbar Section, Showing Vertebrae, Spinal Cord, and Intestines

6. Posterior Aspect of Caudal Area

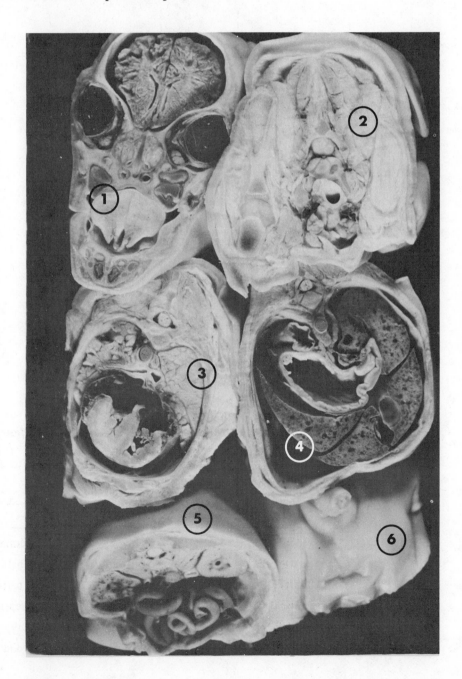

EXERCISE 3   *Terms Designating Abdominal Regions*

The terms below describe abdominal regions (see Figure 3):

1. Right **hypochondriac** region.
2. **Epigastric** region.
3. Left **hypochondriac** region.
4. Right **lumbar** region.
5. **Umbilical** region.
6. Left **lumbar** region.
7. Right **iliac** region.
8. **Hypogastric** region.
9. Left **iliac** region.

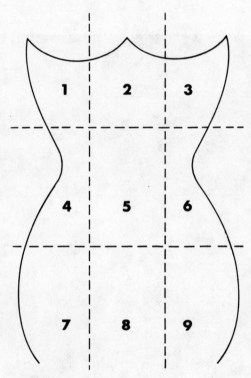

FIGURE 3   *Abdominal regions*

EXERCISE 4   *Prefixes Indicating Location, Direction, and Tendency*

The prefixes given in the table below indicate location, direction, and tendency.

Prefix	Meaning	Example
AB-	From, away	Abnormal—away from normal
AD-	To, near, toward	Adrenal—adjoining the kidney
ANTE-	Before	Antepartum—before delivery
ANTI-	Against	Antiseptic—agent against infection
CIRCUM-	Around	Circumocular—around the eye
CO-	With, together	Coordination—to work together

Prefix	Meaning	Example
CON-	With, together	Congenital—with birth
CONTRA-	Against	Contraindicated—not indicated
COUNTER-	Against	Counterirritant—nonirritating
DIS-	Apart from	Disarticulation—taking a joint apart
ECT-	Outside	Ectonuclear—outside the nucleus of a cell
END-	Within	Endocardium—membrane lining the inside of the heart
EPI-	Upon	Epidermis—upon the dermal layer of skin
EX-	Out from	Exhale—to breathe out
HYPO-	Under	Hypodermic—under the skin
IM-	Not	Immature—not mature
IN-	Not	Incurable—not curable
INFRA-	Under	Infrapatellar—under the kneecap
PERI-	Around	Pericardium—membrane around the heart
POST-	After	Postmortem—after death
PRE-	Before	Prenatal—before birth
PRO-	Before	Prognosis—a fore-knowing
SUPER-	Above	Superciliary—above the eyebrow
SUPRA-	Above	Suprapubic—above the pubic bone
SYM-	With, together	Symphysis—a growing together
SYN-	With, together	Synarthrosis—a bony union
TRANS-	Through	Transurethral—through the urethra

EXERCISE 5   *Prefixes Denoting Number and Measurement*

The prefixes in the table below denote number and measurement.

Prefix	Meaning	Example
UNI-	One	Unicellular—consisting of one cell
MON-	One	Mononuclear—one nucleus
BI-	Two	Bilateral—affecting two sides
BIN-	Two	Binocular—two-eyed
DI-	Two	Dicephalic—two heads
TER-	Three	Tertiary—the third stage
TRI-	Three	Trilobar—three lobes
QUADR-	Four	Quadriceps femoris—group of four muscles in the thigh

Prefix	Meaning	Example
TETRA-	Four	Tetralogy of Fallot—heart anomaly having four features
POLY-	Many	Polydactyly—having (abnormally) many digits
MACR-	Large	Macrocephalic—having an exceptionally large head
MEGA-	Great	Megadontia—exceptionally large teeth
MICRO-	Small	Microscope—an instrument for viewing small things
OLIGO-	Few	Oligophagous—eating only a few kinds of food

EXERCISE 6 *Prefixes Denoting Organs and Structures*

The prefixes in the table below refer to organs and structures of the body.

Prefix	Meaning	Example
ABDOMIN/O-	Abdomen	Abdominal—pertaining to the abdomen
ACR/O-	Extremity	Acromegaly—unusually large extremities
ADEN/O-	Gland	Adenitis—inflammation of a gland
ANGI/O-	Vessel	Angiogram—visualization of blood vessels
ARTHR/O-	Joint	Arthritis—inflammation of a joint
CARDI/O-	Heart	Cardiology—study of the heart
CHONDR/O-	Cartilage	Chondroma—a cartilaginous tumor
CYST/O-	Bladder	Cystoscopy—examination of the inside of the bladder
CYT/O-	Cell	Cytokinesis—cytoplasmic division
DENT/O-	Tooth	Dental—referring to the teeth
DERMAT/O-	Skin	Dermatologist—physician specializing in skin diseases
DERM/O-	Skin	Dermatitis—inflammation of the skin
DUODEN/O-	Duodenum	Duodenal—having to do with the duodenum (first portion of the small intestine)
GASTR/O-	Stomach	Gastrointestinal—having to do with the stomach and intestines

Prefix	Meaning	Example
HEPAT/O-	Liver	Hepatitis—inflammation of the liver
LARYN/GO-	Larynx	Laryngoscope—instrument for viewing inside the larynx
MY/O-	Muscle	Myocardium—heart muscle
NEPHR/O-	Kidney	Nephrology—study of the kidneys
NEUR/O-	Nerve	Neurologist—a physician specializing in diseases of the nervous system
OSTE/O-	Bone	Osteocyte—bone cell
OT/O-	Ear	Otology—study of the ear
PATH/O-	Disease	Pathological—relating to disease
PNEUMON/O-	Lung	Pneumonia—inflammation of the lung
RHIN/O-	Nose	Rhinitis—inflammation of the nasal passages
STOMAT/O-	Mouth	Stomatitis—inflammation of the mouth
THORAC/O-	Thorax or chest	Thoracentesis—puncture of the thorax for the removal of fluid

EXERCISE 7 *Suffixes Denoting Relations, Conditions, and Agents*

The suffixes below denote relations, conditions, and agents.

Suffix	Meaning	Example
-AC	Related to	Cardiac—relating to the heart
-IOUS	Related to	Contagious—communicable by touch
-IC	Related to	Pyloric—relating to the pyloric valve
-ISM	Condition	Mutism—muteness
-OSIS	Condition	Tuberculosis—infection by tuberculosis bacteria
-TION	Condition	Constipation—condition of being constipated
-IST	Agent (one who practices)	Allergist
-OR	Agent	Operator
-ER	Agent	Examiner
-ICIAN	Agent	Physician

EXERCISE 8   *Suffixes Used in Operative Terminology*

The suffixes below are used in operative terminology.

Suffix	Meaning	Example
-CENTESIS	To puncture	Amniocentesis—puncture of the amniotic sac
-ECTOMY	To cut out or excise	Appendectomy—excision of appendix
-OSTOMY	To cut into to form an opening	Colostomy—opening in the large intestine
-OTOMY	To cut into	Tracheotomy—cut into the trachea
-PEXY	To fix or repair	Gastropexy—repair of the stomach
-PLASTY	To repair or reform	Rhinoplasty—repair of the nose
-(R) RHAPHY	To suture (a seam)	Arteriorrhaphy—suture of an artery
-SCOPY	To view	Otoscope—an instrument used to view in the ear

EXERCISE 9   *Miscellaneous Suffixes*

Suffix	Meaning	Example
-ALGIA	Pain	Neuralgia—nerve pain
-EMIA	Of the blood	Viralemia—viruses in the blood
-GRAM	Writing	Electrocardiogram—tracing of the electrical activity of the heart
-ITIS	Inflammation of	Appendicitis—inflammation of the appendix
-OLOGY	Study of	Ophthalmology—study of the eye
-ORRHEA	Flow	Amenorrhea—cessation of menstrual flow
-PHOBIA	Fear of	Claustrophobia—fear of confined spaces

Medical Terminology

DISCUSSION

1. Complete these statements which refer to human body directions, planes, and abdominal regions:

   a. The head is at the _____*Superior*_____ end of the body.

   b. The knee is _____*Inferior*_____ to the hip and _____*Superior*_____ to the ankle.

   c. In anatomical position, the thumb would be _____*Lateral*_____ to the little finger.

   d. The vertebral column is _____*Dorsal*_____ to the stomach.

   e. When a surgeon amputates a leg, he makes a _____*Transverse*_____ cut through the bone.

   f. The liver is located primarily in the _____*Rt. hypochondriac*_____ region of the abdomen.

2. Using the prefixes and suffixes in this unit, what would be the meaning of these terms?

   a. unilateral _____*one*_____

   b. periosteum _____*around the bone*_____

   c. interdigital _____*between the digits*_____

   d. pathologist _____*person who studies disease*_____

   e. pneumocentesis _____*to puncture the lung*_____

   f. tonsillectomy _____*cut out the tonsils*_____

   g. gastrotomy _____*to cut into the stomach*_____

   h. lymphoma _____

3. What would be a medical term for each of the following?

   a. Suturing a kidney _____*nephrorrhaphy*_____

   b. Inflammation of a joint _____*arthritis*_____

   c. Viewing the interior of a bronchus _____*ostomy*_____

   d. Pain in a muscle _____*myalgia*_____

UNIT II   *The Microscope*

PURPOSE   The purpose of Unit II is to familiarize the student with the structure and function of the microscope in order to better understand cytology and histology.

OBJECTIVES   In order to complete Unit II, the student must be able to do the following:

1. Name and identify the major parts of the microscope.
2. State and demonstrate the functions of the microscope parts.
3. List and follow the directions for proper care of the microscope.
4. Demonstrate use of the microscope.

MATERIALS   compound microscope   prepared slides of newsprint and
              unprepared slides   other specimens

PROCEDURE

EXERCISE 1   *The Structure of the Compound Microscope*

The microscope is an invaluable tool in your study of anatomy and physiology. Through its proper use you will be able to see very small structures, such as cells, which contribute to the total structure and function of living organisms. In order that you maximize your benefits from the use of the microscope it is essential that you know the basic facts about microscopy. You must know the parts of the microscope and how to use each of its components.

The compound microscope is commonly used in laboratory studies and consists of two lenses or lens systems. They are the ocular and objective. The ocular or eyepiece magnifies an object which is further magnified by the objective. Total magnification is the product of ocular magnification × objective magnification.

Even though you may have used a microscope previously, work through this exercise so that you ensure its efficient use.

Your instructor will assign a microscope to you. You are to use this same microscope for your studies throughout this course.

Obtain the microscope assigned to you. Always observe the following when carrying the microscope to your work area:

1. Put one hand beneath the scope for support.
2. Put the other hand around the curved arm.
3. Carry the microscope in an upright position.

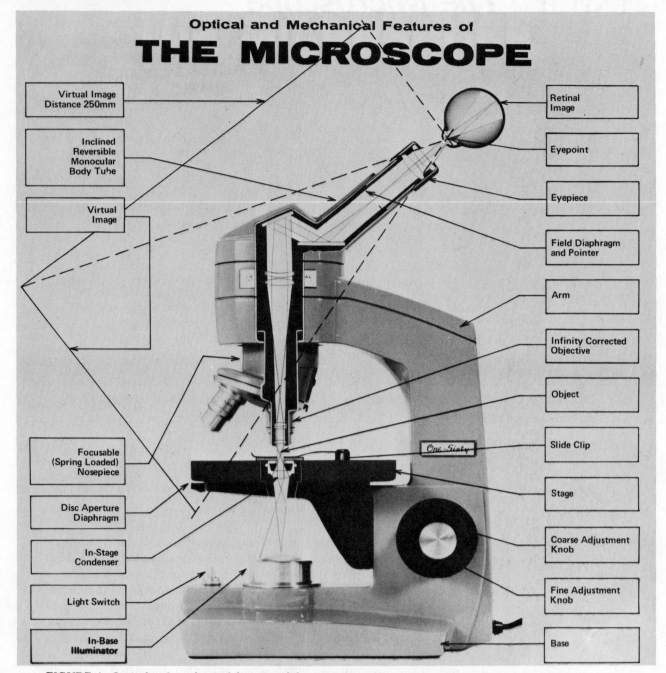

FIGURE 4 *Optical and mechanical features of the microscope* (Courtesy American Optical Corporation)

Now place the scope in front of you at your work area and insert the cord into the electrical outlet.

Using Figure 4, identify the parts of the microscope listed below:

1. Ocular or eyepiece.
2. Arm.
3. Revolving nosepiece with objectives.
4. Coarse adjustment.
5. Fine adjustment.
6. Specimen stage with slide holder.
7. Condenser.
8. Diaphragm.
9. Base with illuminator.

EXERCISE 2   *Functions of the Microscope Parts*

Reexamine the microscope parts which you identified in Exercise 1, noting the function of each part. Keep in mind that your primary purpose is to adjust the microscope so that you get the desired magnification and clarity (resolution) of the object viewed. Especially note the parts which allow you to achieve your primary purpose in focusing.

MICROSCOPE PARTS AND THEIR FUNCTIONS

Parts	Function
1. Ocular	A lens of a given magnification which is probably engraved on the rim, e.g., 10×. (You may also see a pointer embedded in the ocular. This is used to aid you in indicating specific locations within the microscopic field.)
2. Revolving nosepiece	This plate, which is capable of rotation, allows you to utilize objectives of different magnifications.
3. Objectives	Lenses of varying magnifications. The values (e.g., 10×, 43×, and 100×) are usually engraved on the objectives.
a. Low power objective	Lens with the least magnification.
b. High power objective	Lens with greater magnification than low power objective.
c. Oil immersion objective	Lens with greatest magnification. (You will infrequently use this lens in this course. Proper technique must be demonstrated by your instructor prior to its use.)
4. Specimen stage with slide holder	Platform upon which the slide is positioned for focusing. Note that your scope may have clips for anchoring the slide or it may have a mechanical slide holder. Obtain a clear slide and fit it into the holder. The slide can be moved forward, backward, and from side to side by rotating the small knob on the underside or on top of the stage.
5. Arm	Handle for holding and positioning microscope.

Parts	Function
6. Coarse adjustment	Larger knobs on both sides of the base of the arm which allow for initial focusing of the object to be viewed.
7. Fine adjustment	Smaller knobs on both sides of the base of the arm which allow for refinement of detail in focusing.
8. Condenser	A lens system which concentrates light from the illumination source so that a cone of light fills the aperture of the objective. After checking to see that the light bulb is on, move the condenser up and down and note the varying intensities of light visible in the ocular.
9. Diaphragm	Plate with apertures (openings) of varying diameters from small to large. Open and close the diaphragm by adjusting it with its handle so that varying intensities of light are visible in the ocular.
10. Base with illuminator	Platform upon which microscope is structured, containing an electric light source.

EXERCISE 3   *Proper Care of the Microscope*

Care must be exercised in removing the microscope from its storage area and carrying it to your work area (see Exercise 1). Care must also be exercised in cleaning the lenses and returning the microscope to its storage area.

In cleaning the lenses, do not remove them. Gently wipe only the external surface of the lenses with lens paper. Do not use any other material for wiping since the lenses are damaged easily.

When returning the microscope to its storage area, make sure to:

1. Put the low power objective down toward the specimen stage.
2. Remove all slides from the stage.
3. Rewind the electrical cord.

EXERCISE 4   *Use of the Compound Microscope*

1. Using a prepared slide of newsprint, you will practice focusing. Obtain the slide from the slide tray and clean it if necessary. Then place the

slide in the slide holder on the specimen stage. Make sure that the slide is properly anchored and movable in the slide holder.

2. Make sure that the light source is on, then adjust the diaphragm so that by looking into the ocular you see a bright field of light. You may also adjust the condenser to increase the intensity of light if necessary. Keep in mind that the intensity and brightness of light must be varied in relationship to the density of the specimen to be focused and the magnification used. Light must be transmitted through the specimen in order to be seen. Therefore, thinner specimens require less light than do thicker specimens.

3. Now make sure the *low* power objective is in place on the revolving nosepiece. It should be in a downward position toward the stage. By watching on the side of the stage and by rotating the coarse adjustment slowly away from you, you can bring the low power objective down to its limit.

4. With both eyes open, look into the ocular. At first you may be distracted with both eyes open but eventually you will become unaware of the surroundings. Only what is being focused will be seen. *Slowly* rotate the coarse adjustment toward you until you see a letter of newsprint. Adjust the slide position on the specimen stage if necessary.

5. Now focus with the fine adjustment knob. The edges of the letter of newsprint should be sharp and well defined. It may be necessary to readjust the light allowing for more or less. Use the condenser and diaphragm for this adjustment.

6. In order to obtain greater magnification, it will now be necessary to rotate the nosepiece and place the high power objective in a downward position. Listen for the clicking sound to make sure it is in position. If your microscope is parfocal the image under high power should be focused immediately. Many classroom microscopes, because of constant use, are not parfocal. In such cases, minor adjustments with both the coarse and fine adjustments must be made.

7. Under high power magnification, more light is often necessary. As the power of the objective increases, the diameter of the lower lens decreases, and less light enters the objective. Again readjust the condenser and diaphragm if necessary.

8. When using high power magnification, never rotate the coarse adjustment while looking into the ocular. Move your head so that you are looking at the side of the stage. Then slowly rotate the coarse adjustment until the objective is close to but *never* touching the slide.

9. *Repeat* the preceding procedure using another demonstration slide from the microscope tray provided by your instructor.

10. Draw and label the images that you have viewed under both low and high power magnification in the spaces provided.

Newsprint Low power magnification	Newsprint High power magnification
Prepared slide Low power magnification	Prepared slide High power magnification

12. Using the guidelines set forth in Exercise 3, clean the microscope and replace it in the storage area. Replace the slides that you have used to their proper position in the slide trays.

The Microscope

DISCUSSION

1. What was the total magnification of the newsprint under low power? _____ under high power? _____

2. What is the magnification of each of the following?

   a. ocular _____

   b. low power objective _____

   c. high power objective _____

3. If your field of magnification under low power appears too dark, what should you do to lighten the field?

4. When using the high power objective, in which direction do you rotate the fine adjustment in order to obtain perfect focusing?

5. What are the two lens systems of a compound microscope?

6. What are the guidelines to be followed when carrying the microscope?

7. What are the guidelines to be followed when returning the microscope to its storage area?

8. When viewing an image under high power, is more or less light necessary? Why?

9. What is meant by parfocal?

10. Does a diaphragm aperture of small diameter clarify or distort an image focused under high power? Explain.

UNIT III *Cells*

A. Cell Structure

PURPOSE
: The purpose of Unit III-A is to acquaint the student with cell structure and cell division.

OBJECTIVES
: To complete Unit III-A, the student must be able to do the following:
  1. Name and identify the organelles found in a **cell**.
  2. Prepare a smear of buccal mucosa and identify the **nucleus**, **cytoplasm**, and **plasma membrane** (cell membrane).
  3. Recognize the stages of **mitosis**.

MATERIALS
: clean microscope slides   flat toothpicks   prepared onion root tip slides
  medicine droppers   methylene blue   prepared whitefish mitosis slides

PROCEDURE

EXERCISE 1   *Animal Cell Structure*

Study the illustration of an animal cell, as seen under an electron microscope (shown in Figure 5):

FIGURE 5 *Generalized animal cell drawn after electron micrograph (Adapted from Biology Today, © 1972 by Ziff-Davis Publishing Company. Courtesy of CRM Books, a division of Random House, Inc.)*

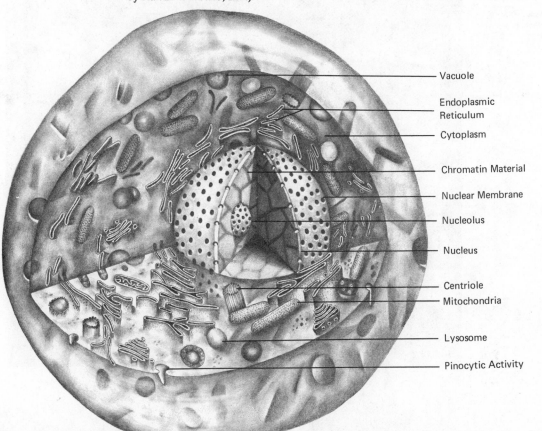

Vacuole

Endoplasmic Reticulum

Cytoplasm

Chromatin Material

Nuclear Membrane

Nucleolus

Nucleus

Centriole

Mitochondria

Lysosome

Pinocytic Activity

EXERCISE 2   *Observation of Buccal Mucosa Cells*

Place a drop of water on a clean slide. Scrape the inside of your check (buccal mucosa) with a flat toothpick and mix with the water on the slide. The thinner the *smear* the better. Allow the slide to air dry. Add one small drop of methylene blue to your smear and cover with a coverslip. Observe under a microscope and draw what you see. *Label:* **nucleus, cytoplasm, plasma membrane.**

EXERCISE 3   *Mitosis*

Obtain an onion root tip; locate the distal portion of the root tip under low power; then switch to high power. Look for the following stages of mitosis:

1. **prophase.**
2. **metaphase.**
3. **anaphase.**
4. **telophase.**

When a cell is not actively dividing, it is in **interphase**. Notice that in interphase **chromatin** is not condensed into chromosomes.

FIGURE 6A *Mitosis in onion root cells*

1. telophase
2. Interphase, Resting
3. Prophase

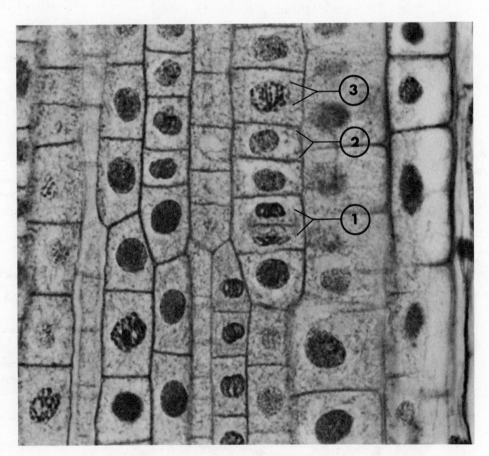

FIGURE 6B *Mitosis in onion root cells*

1. Daughter cells
2. Telophase

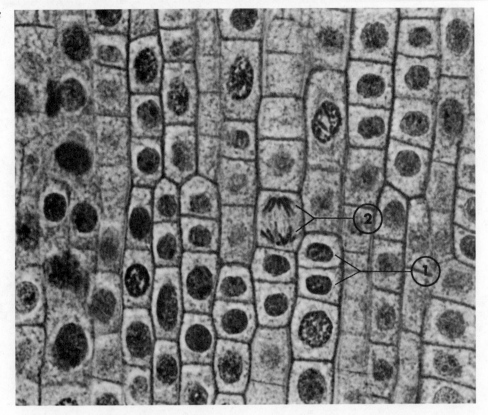

FIGURE 6C *Mitosis in onion root cells*

1. Metaphase
2. Anaphase
3. Prophase, Late

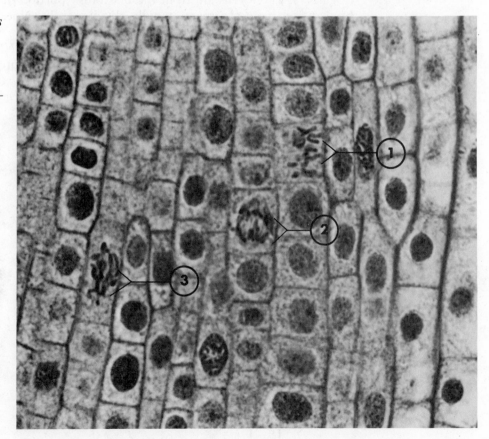

FIGURE 7
*Mitotic figures in
whitefish eggs*

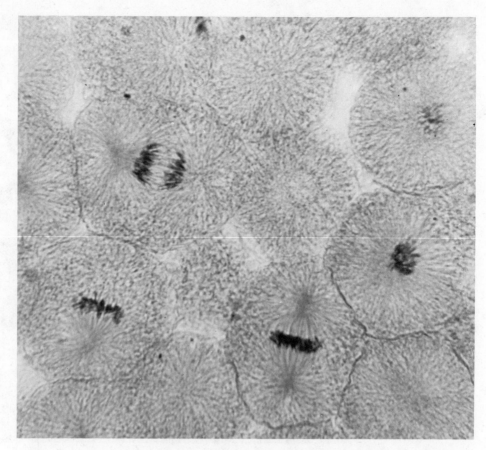

Observe a whitefish mitosis slide and look for similar mitotic figures
(Figure 7). These cells are from the blastodisk portion of the whitefish embryo, which contains many rapidly dividing cells. Note that animal cells
lack a cell wall.

Observe a prepared slide of human chromosomes. Note the difference
between XY (male) and XX (female) chromosomes (Figure 8).

FIGURE 8 *Normal
human somatic
chromosomes*
(Courtesy Carolina
Biological Supply
Company)

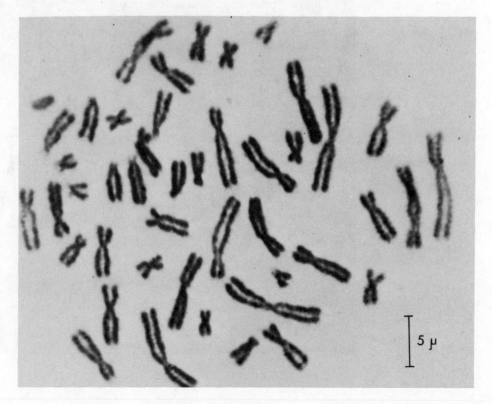

5 µ

B. Cell Physiology

PURPOSE	The purpose of Unit III-B is to give the student an understanding of some basic concepts of cell physiology.

OBJECTIVES	To complete Unit III-B, the student must be able to do the following:

1. Observe various types of diffusion.
2. Demonstrate the principles of osmosis and dialysis.
3. Define a hypertonic, an isotonic, and a hypotonic solution.

MATERIALS

ammonia or ether
distilled water
potassium permanganate crystals
test tubes containing gelatin
probes
sodium chloride (NaCl)
glucose
raw egg white or albumin
molasses
large white potato that has been
   soaked overnight
½- to ¾-in. diameter drill bit

1-ml and 5-ml pipettes
dialysis membrane
blood
test tubes and racks
dilute nitric acid
1% silver nitrate solution
1.5% saline solution
Benedict's solution
large beakers
capillary tubing, 60 cm long
rubber stopper
ring stands and test tube clamps

PROCEDURE

EXERCISE 1   *Diffusion*

There are three major types of diffusion:

1. *Diffusion of gases.* Open a bottle of ammonia or ether at the front of the room. Why may the odor soon be detected in all parts of the room?

2. *Diffusion within a liquid.* Drop a crystal of potassium permanganate ($KMnO_4$) into a beaker of water. Observe during the laboratory period. What happens?

Slowly the potassium permanganate diffuses evenly through the water.

#4

3. *Diffusion through a colloid.* Using a straight probe and medicine dropper, stab a tube of gelatin and add one to two drops of methylene blue. Observe during the laboratory period and at the beginning of the next period. Result?

EXERCISE 2   *Osmosis*

Bore a hole approximately two-thirds into the length of a large white potato using a drill bit. This exercise will give better results if the potato has been immersed in water overnight or for at least 8 hours. Fill the hole to within ¾ to 1 in. of the top with molasses. Insert a piece of glass tubing (capillary tubing is preferable) into a rubber stopper, and insert stopper into the potato containing the molasses. The tubing should extend nearly to the bottom of the hole.

Place the prepared potato into a large beaker nearly full of water and place on a ring stand, anchoring the tubing with a test tube clamp. It is a good idea to wrap paper toweling around the tubing at the point where the clamp will be fastened to hold it more tightly.

Observe the apparatus during the laboratory period. The molasses should rise into the tubing as a result of water passing through the potato.*

EXERCISE 3   *Dialysis*

Place water, sodium chloride, glucose, and albumin or raw egg white from one egg into a bag of dialysis membrane or cellophane and suspend from a rod across a beaker of water. Allow the bag to be immersed in the water. Let stand for 1 to 2 hours; then run the following tests on the water in the beaker and solution in the dialysis bag.

1. *Test for albumin.* Pour about 5 ml of fluid from the beaker into a test tube and add a few drops of nitric acid (HNO_3). If albumin is present, it will be coagulated by the HNO_3 and turn white. Conclusion?

Albumin was not present because the liquid remained clear. Albumin did not pass through the membrane. Liquid

*The authors wish to thank Dr. Calvin Kuehner for contributing this exercise.

2. *Test for NaCl.* Pour about 5 ml of fluid from the beaker into another test tube and add a drop of silver nitrate ($AgNO_3$). If the solution turns cloudy or white, silver chloride (AgCl) has been formed and the test is positive. Conclusion?

NaCl was present
more sodium was on the outside of
the membrane than in
Inside was slightly ch...
Outside was white

3. *Test for glucose.* Put 5 ml of Benedict's solution into a test tube and add four to five drops of the beaker water. Boil 2 minutes. Cool slowly. If a green, yellow, or red precipitate forms, the presence of glucose is indicated. Conclusion?

Glucose did not diffuse through the dialysis bag.

Repeat the above tests using 5-ml aliquots of solution from within the dialysis bag.

EXERCISE 4

Hypertonic, Isotonic, and Hypotonic Solutions

If you are to understand this exercise, you should become familiar with the following concepts:

1. If cells are placed in a **hypertonic** solution, in which the solute concentration is greater outside the cell than inside, the cells will lose solvent (usually water) and shrivel. In erythrocytes (red blood cells), this process is known as *crenation.*
2. If cells are placed in an **isotonic** solution, in which the solute concentration outside is the same as inside, there will be no net change in the amount of water in the cell. Therefore, the cells will retain their original shape.
3. If cells are placed in a **hypotonic** solution, in which the solute concentration is greater inside the cell than outside, there will be a net gain of water in the cell, causing a rise in intracellular pressure, and the cells will burst. In erythrocytes, this is known as *hemolysis.*

Set six small test tubes into a rack and number left to right 1, 2, 3, 4, 5, and 6. First, pipette 1.5% saline into the tubes in the amounts shown in the table below. Then pipette the indicated amounts of water into each tube.

Test Tube No.	Add 1.5% Saline in Following Amounts (ml)	Add Distilled Water in Following Amounts (ml)	Final Concentration of Saline (%)
1	0	5.0	0
2	1.0	4.0	0.3
3	2.0	3.0	0.6
4	3.0	2.0	0.9
5	4.0	1.0	1.2
6	5.0	0	1.5

Carefully pipette 0.2 ml of blood into each tube and mix gently by tilting and rotating. Mix gently and thoroughly. In the space below, describe the appearance of the tubes 5 minutes after the blood was added and mixed.

Tube 1:

Tube 2:

Tube 3:

Tube 4:

Tube 5:

Tube 6:

Cells

DISCUSSION

1. Using your textbook or other references, state the function of each of these cellular organelles:

   a. chromatin _____

   b. nucleolus _____

   c. centrioles _____

   d. plasma membrane _____

   e. endoplasmic reticulum _____

   f. ribosomes _____

   g Golgi apparatus _____

   h. lysosomes _____

   i. mitochondria _____

   j. nuclear membrane _____

2. Using your textbook or other references, name the stage(s) of mitosis during which each of the following occurs in the cell:

   a. cell is not actively dividing _____

   b. cytokinesis occurs _____

   c. chromosomes are lined up at the equator of the cell _____

   _____

   d. chromatin condenses into chromosomes _____

   _____

   e. nuclear membrane is not present _____

   f. DNA is replicating _____

3. What physical principle determines whether various types of molecules will pass through dialysis membrane?

   _____

   _____

4. Write the chemical equation for the reaction between sodium chloride (NaCl) and silver nitrate ($AgNO_3$).

   _____

5. From your results in Exercise 4, which approximate final concentration of saline is isotonic to erythrocytes?

   _____

UNIT IV *Tissues*

PURPOSE   The purpose of Unit IV is to enable the student to recognize the various tissue types.

OBJECTIVES   In order to complete Unit IV, the student must be able to do the following:

1. Name the general categories of tissues found in the body.
2. Identify general and specialized tissue types microscopically.
3. State the location of various tissue types in the body.
4. Identify the epidermal and dermal layers of skin.
5. Microscopically differentiate between Caucasian and Negroid skin.

MATERIALS   Slides of the following tissue types:

simple squamous epithelium	dense fibrous connective tissue
simple cuboidal epithelium	areolar tissue
simple columnar epithelium	adipose tissue
stratified squamous epithelium	hyaline cartilage
skin—Negroid and Caucasian	osseous tissue
pseudostratified ciliated columnar	skeletal muscle
epithelium	smooth muscle
transitional epithelium	cardiac muscle

Tissues are groups of cells that perform a common function. There are four general tissue types. They include *epithelial* tissue, which covers a surface and functions in protection, secretion, and absorption; *connective* tissue, which binds and supports; *muscle* tissue, which contracts; and *nervous* tissue, which conducts impulses.

These four basic tissue types can be further subdivided. Epithelial tissue (**epithelium**) is comprised of many cells and very little intercellular material. Epithelium can be **simple**, meaning one cell layer thick, or **stratified**, which indicates several cell layers. If the cells comprising the epithelium are flat, they are known as **squamous**; if they are cube-shaped, they are **cuboidal**; and if they are taller than they are wide, the epithelium is referred to as **columnar**. Occasionally, the nuclei of a single cell layer are at different levels, giving the appearance of stratified epithelium. This type is referred to as **pseudostratified**. Another epithelial type consists of stratified balloon-shaped cells and possesses the property of stretching. This is known as **transitional** epithelium and is found in parts of the urinary tract.

Connective tissue, in contrast to epithelial tissue, contains relatively few cells and much intercellular material. It may be classified as **dense** and **fibrous**, as found in tendons and ligaments, or **loose, areolar**, which forms the "packing" in and around organs. There are several specialized types of connective tissue. **Adipose**, or fat, consists of cells in which most of each cell's volume is occupied by fat, rather than cytoplasm. This feature imparts a characteristic "signet ring" appearance to adipose cells. Cartilage and bone are other specialized connective tissue types. Cartilage may be **hyaline, elastic,** or **fibrous.** Bone, or **osseous** tissue, may be *compact* or *spongy.* Most biologists classify *blood* as a specialized connective tissue, while others consider it a fifth basic tissue type. You will be studying blood in Unit VIII.

There are three types of muscle tissue: **skeletal, cardiac,** and **smooth.** The primary functional cells in nervous tissue are known as **neurons.** Muscle and nerve slides will be studied superficially in this unit and in greater depth in later exercises.

Using the following illustrations, examine the slides of tissue types your instructor has made available to you. By studying the slides and the summary table, you should be able to answer most of the questions in the discussion.

PROCEDURE

EXERCISE 1   *Microscopic Identification of Tissue Types*

Examine each of the slides and identify each tissue type using the illustrations in this unit as a guide (Figures 9–26):

FIGURE 9 *Human squamous epithelial cells, smear from mouth*

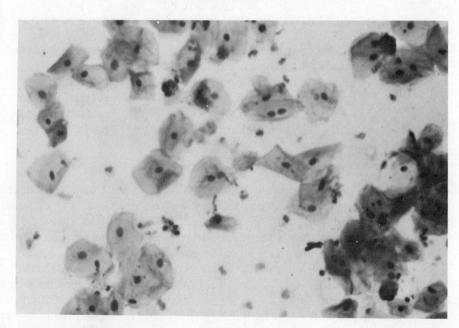

FIGURE 10 *Simple columnar epithelium*

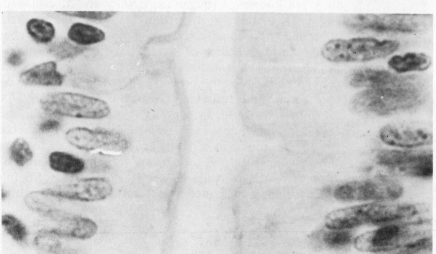

FIGURE 11 *Ciliated columnar epithelium*

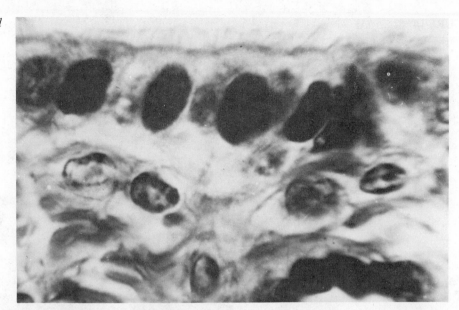

FIGURE 12 *Cuboidal epithelium as found in the collecting tubules of the kidney*

1. Cuboidal Cells

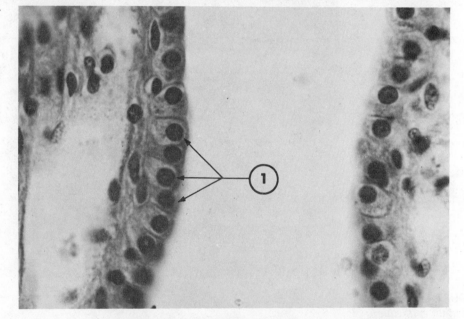

FIGURE 13 *Ciliated pseudostratified columnar epithelium*

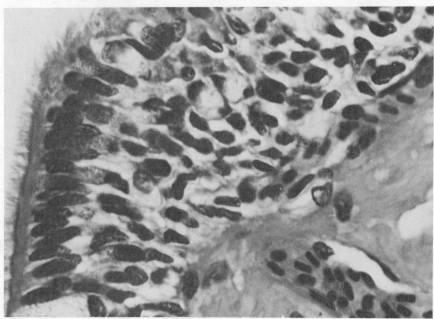

FIGURE 14 *Stratified squamous epithelium*

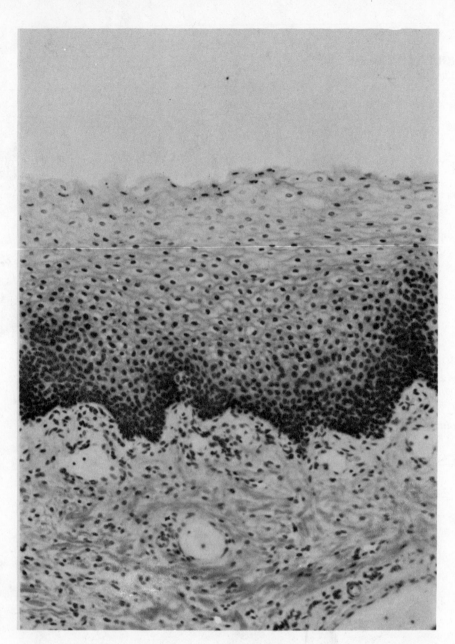

FIGURE 15 *Stratified columnar epithelium*

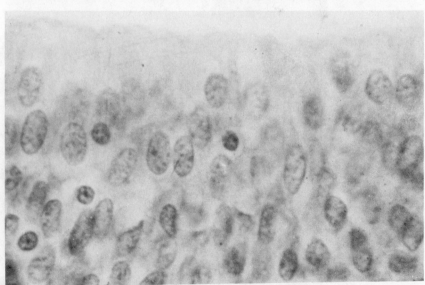

FIGURE 16 *Transitional epithelium*

1. Balloon-shaped cells next to the lumen

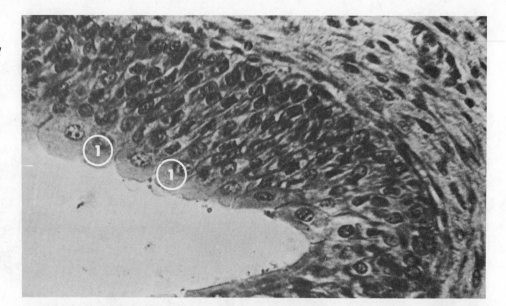

FIGURE 17 *Smooth muscle fibers, longitudinal section*

1. Fiber
2. Nucleus

FIGURE 18 *Smooth muscle*

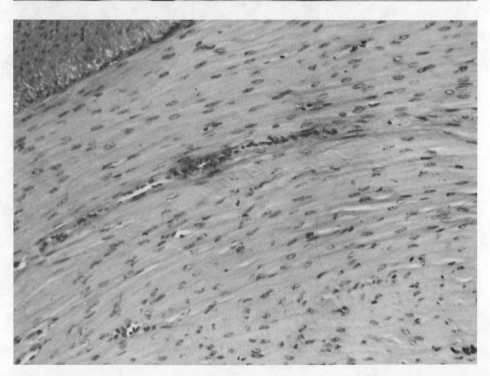

FIGURE 19 *Striated muscle fiber, longitudinal section*

1. Nuclei

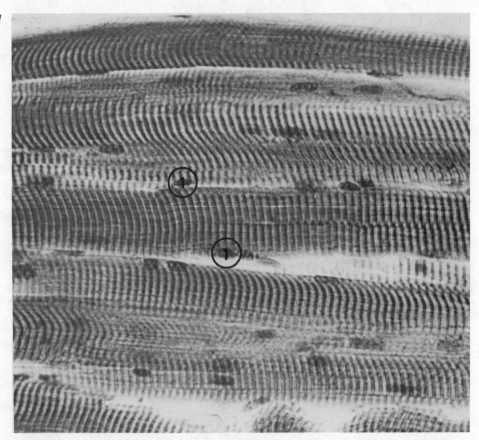

FIGURE 20 *Cardiac muscle, longitudinal section*

1. Nucleus
2. Intercalated disks
3. Capillary with blood cells

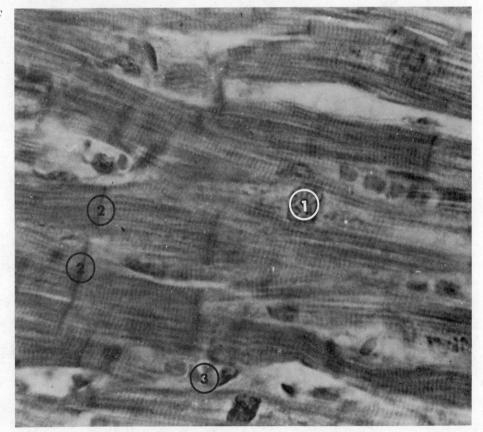

FIGURE 21 *Elastic cartilage*

1. Chondrocyte Cells
2. Interterritorial Matrix
3. Perichondrium

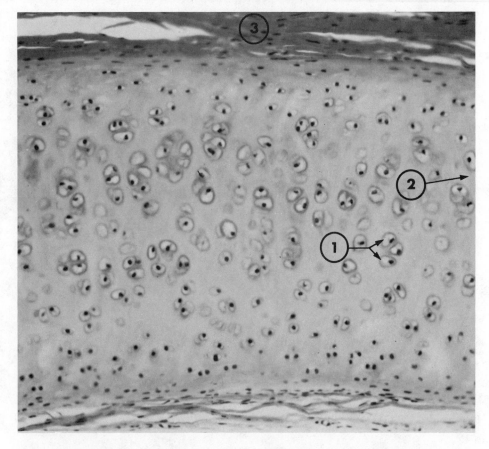

FIGURE 22 *Hyaline cartilage of the trachea*

1. Lacuna
2. Elastic fibers
3. Chondrocyte

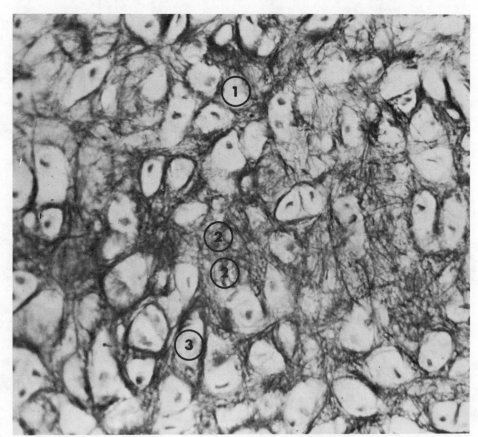

FIGURE 23
*Ground bone show-
ing Haversian sys-
tems*

1. Bone matrix
2. Haversian canal
3. Osteocytes

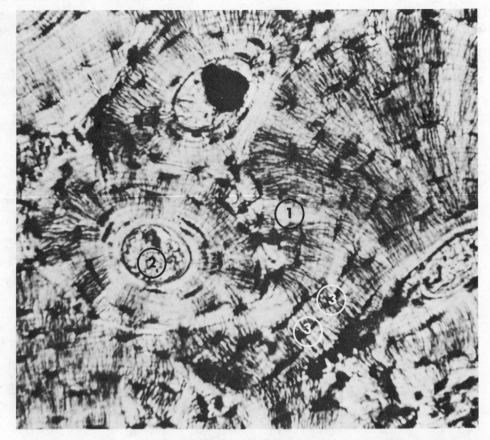

FIGURE 24 *Long
bone developing*

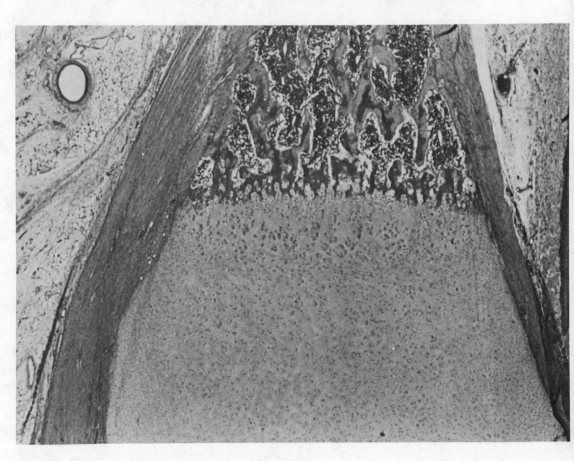

FIGURE 25 *Adipose tissue*

1. Signet-ring-
   shaped cell
2. Fat

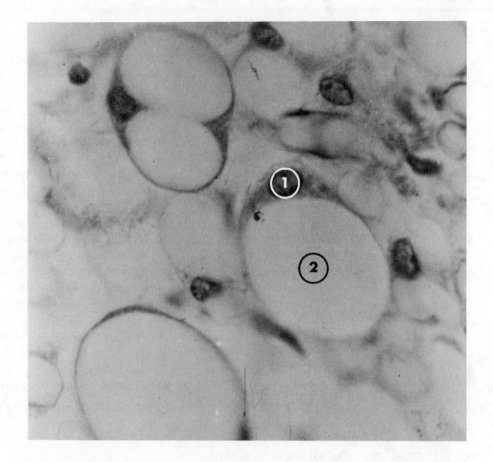

FIGURE 26 *Loose, irregular connective tissue*

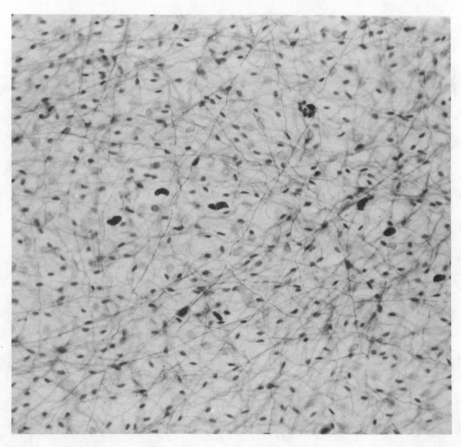

EXERCISE 2   *Microscopic Identification of the Skin*

The skin is comprised of an outer **epidermis** and a deeper layer, the **dermis** or **corium**. The epidermis is composed of stratified squamous epithelium and consists of five layers:

1. **Stratum corneum.** This is the outermost layer of the epidermis, consisting of flattened, dead, keratinized (converted to protein) cells that are continuously shed. The stratum corneum helps serve as a barrier to light, heat, bacteria, and most chemicals.
2. **Stratum lucidum.** This layer is directly beneath the stratum corneum and is one or two cell layers thick. It may be difficult to see on your microscope slide.
3. **Stratum granulosum.** This is a thin layer lying beneath the stratum lucidum. It is thought to be the layer in which keratinization takes place. In this layer granules are numerous and tend to stain heavily.
4. **Stratum spinosum.** This layer may also be difficult to see. It is beneath the stratum granulosum and consists of "prickly" cells.
5. **Stratum germinativum.** This layer is the deepest layer of the epidermis. Cells in this layer actively undergo mitotic division and give rise to the four outer epidermal layers. **Melanin**, the principal pigment of the skin, is formed in this layer.

The dermis lies beneath the epidermis. It contains connective tissue fibers, blood vessels, nerves, sweat glands, sebaceous glands, and hair follicles.

Examine slides of Caucasian and Negroid skin, identifying the above structures (Figures 27–29).

FIGURE 27 *Thick skin, palm, showing epidermal layers*

1. Stratum Corneum   3. _____   5. Dermal Papillae

2. Stratum Lucidum   4. Stratum Germinativum   6. Epidermis

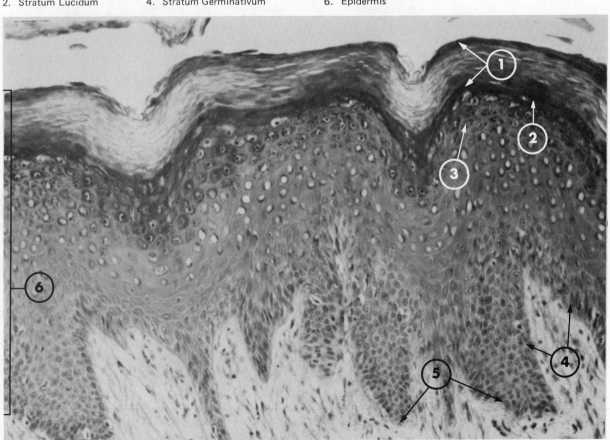

FIGURE 28 *Skin, human, showing excretory duct of a sweat gland*

1. Stratum Corneum

2. Excretory Duct of Sweat Gland

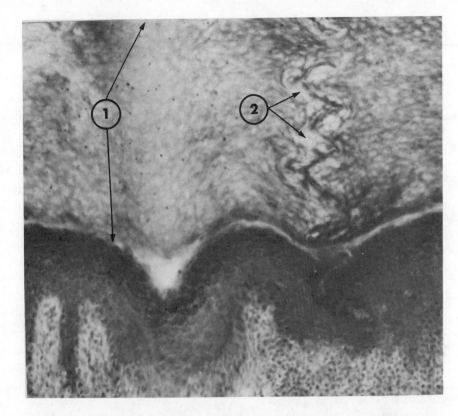

FIGURE 29 *Skin, human scalp, showing hair follicles*

1. _____

2. _____

3. Hair Follicle

4. _____

5. _____

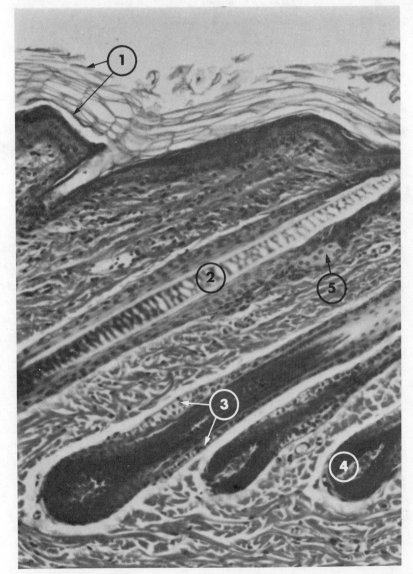

CLASSIFICATION OF TISSUES

Primary Tissue	Types		Divisions		Example
Epithelium	A. Covering external body surface or lining internal surface	Simple	Squamous		Bowman's capsule (kidney)
			Cuboidal		Collecting tubule (kidney)
			Columnar		Gallbladder (nonciliated)
					Uterine tube (ciliated)
					Intestinal mucosa
		Pseudo-stratified	Columnar		Male urethra (nonciliated)
					Trachea (ciliated)
		Stratified	Squamous		Skin (keratinizing)
					Vagina (nonkeratinizing)
					Cornea
			Cuboidal		Sweat glands
			Columnar		Male urethra
			Transitional		Urinary bladder
	B. Multicellular glands	Exocrine	Simple		Gastric, sweat
			Compound		Salivary
		Endocrine			Thyroid, adrenal
Muscle	A. Smooth (involuntary)				Intestinal tract, blood vessels
	B. Striated (voluntary)				Skeletal muscle
	C. Cardiac (involuntary)				Heart muscle
Connective tissue	A. General	Loose	Mesenchyme		Embryonic and fetal tissue
			Mucoid		Wharton's jelly (umbilical cord)
			Areolar		Found in most organs and tissues
			Adipose		Subcutaneous tissue
			Reticular		Bone marrow, lymph nodes
		Dense	Irregular		Dermis, capsules of organs
			Regular		Tendon, cornea
	B. Special		Cartilage	Hyaline	Costal cartilage, trachea
				Fibrous	Intervertebral disc
				Elastic	External ear, epiglottis
			Bone	Cancellous	Epiphyses of long bones
				Compact	Shaft of long bone
			Hemopoietic	Myeloid	Bone marrow
				Lymphoid	Spleen, lymph node
			Blood		
			Lymph		
Nervous tissue	A. Central nervous system	Gray matter			Brain, spinal cord
		White matter			Brain, spinal cord
	B. Peripheral nervous system	Nerves			
		Ganglia			
		Nerve endings			
	C. Special receptors				Eye, ear, nose

Tissues

DISCUSSION

1. For review, draw the following tissue types in the spaces provided:

Stratified squamous epithelium	Simple columnar epithelium	Adipose	Dense fibrous connective tissue
Hyaline cartilage	Osseous	Skeletal muscle	Smooth muscle

2. a. In which epidermal layer are melanocytes found? _____

    _____

    b. Which layer(s) of epidermis is (are) comprised of dead cells? _____

    _____

    c. Which epidermal layer(s) is (are) keratinized? _____

    _____

3. Which specialized tissue type(s) would you find in the following locations?

    a. collecting tubules of the kidney _____

    b. lining the urinary tract _____

    c. trachea _____

    d. muscle layer of the small and large intestines _____

    e. mesenteries _____

    f. biceps muscle _____

    g. tendon _____

    h. in the nasal septum _____

    i. lining the intestinal tract _____

    j. fat _____

UNIT V   *Skeletal System*

PURPOSE   The purpose of Unit V is to familiarize the student with the various divisions of the human skeleton.

OBJECTIVES   To complete Unit V, the student must be able to do the following:

1. Identify major components of a Haversian system microscopically.
2. Identify major anatomical structures of a long bone.
3. Identify bones of both axial and appendicular skeletons.
4. Identify major bone markings.
5. Identify locations of major sutures and fontanels.
6. Differentiate between diarthrotic and synarthrotic joint.

MATERIALS   articulated human skeleton
slides of dry ground bone
  (osseous tissue)
disarticulated human skeleton

models: sagittal section of a long bone, adult and infant labeled skulls, Beauchene skull models (optional)

PROCEDURE

EXERCISE 1   *Microscopic Examination of Osseous Tissue*

Since you studied osseous tissue previously (Unit IV), this should be a review. Examine a prepared slide of dry ground bone and identify the following: Haversian canal, lamella, lacuna, canaliculi, osteocytes, and interstitial material (see Figure 23 in Unit IV).

  Draw a Haversian system and label the structures listed above to the right of your diagram:

EXERCISE 2   *Examination of the Gross Anatomy of a Long Bone*

You are to identify the gross anatomical features of a sagittal section of a long bone. Examine the available specimen and identify the following: articular cartilage(s), compact bone, cancellous bone, diaphysis, endosteum, epiphysis, medullary canal, and periosteum.

EXERCISE 3   *Identification of Bones of the Human Skeleton*

You are to identify the bones of both the axial and appendicular skeleton. Identify the bones of the axial skeleton first. If you are working with a disarticulated (bones disjointed) skeleton, arrange the bones in the proper linear position on the laboratory table.

There are more than 200 bones in the human adult; some are paired, some are single. Other bones—such as the skull, sternum, and hip—are fusions of a number of bones. Be sure to identify the separate bones that comprise each fusion.

In infancy and early childhood, ossification has not been completed; therefore bones are more cartilaginous. If you have a young skeleton, observe the delicate nature of the bones and obvious cartilage.

Identify the following bones from the chart:

Skeletal Division	*Name of Bone(s)*	*Number of Bones*
A. Axial Skeleton		80 (total)
1. Skull		28
a. Cranium		8
	Frontal	1
	Parietal	2
	Temporal	2
	Occipital	1
	Sphenoid	1
	Ethmoid	1
b. Face		14
	Mandible	1
	Nasal	2
	Lacrimal	2
	Inferior conchae (turbinates)	2
	Zygomatic (malar)	2
	Palatine	2
	Maxillary	2
	Vomer	1
c. Ear Ossicles		6
	Malleus	2
	Incus	2
	Stapes	2
2. Hyoid		1

Skeletal Division	*Name of Bone(s)*	*Number of Bones*	
3. Vertebrae		26	
	Cervical	7	
	Thoracic	12	
	Lumbar	5	
	Sacrum	1	(fusion of 5)
	Coccyx	1	(fusion of 3–5)
4. Sternum		1	(fusion of 3)
	Manubrium	1	
	Gladiolus (body)	1	
	Xiphoid process	1	
5. Ribs		12 pairs (7 pairs true ribs, 5 pairs false ribs)	
B. Appendicular Skeleton		126	(total)
1. Upper Division		64	(total)
	Clavicle	2	
	Scapula	2	
	Humerus	2	
	Radius	2	
	Ulna	2	
	Carpals (*see Figure 47 in Exercise 4 of this unit*)	16	(8 per hand)
	Scaphoid	1	
	Lunate	1	
	Triquetrum	1	
	Pisiform	1	
	Trapezium	1	
	Trapezoid	1	
	Capitate	1	
	Hamate	1	
	Metacarpals	10	(5 per hand)
	Phalanges	28	(14 per hand)
2. Lower Division		62	(total)
	Ossa Coxae	2	(each a fusion of 3)
	Ilium	1	
	Ischium	1	
	Pubis	1	
	Femur	2	
	Patella	2	
	Tibia	2	
	Fibula	2	
	Metatarsals	10	(5 per foot)
	Tarsals (*see Figure 54 in Exercise 4 of this unit*)	14	(7 per foot)
	Talus	1	
	Calcaneus	1	
	Cuboid	1	
	Navicular	1	
	Cuneiforms	3	
	Phalanges	28	(14 per foot)

EXERCISE 4   *Identification of Bone Markings*

Since the skeletal system serves as the basis for body topography, one should know more about the bones than simply their names and locations. Knowledge of *bone markings*, which are specific identifiable projections or depressions on bones, is of great importance for it facilitates identification of other body structures such as muscles, blood vessels, and nerves.

Below are listings of the major bone markings. Those classified as projections (also called **processes**) grow out from the bone, whereas those classified as depressions (also called **fossae**) are indentations in the bone. Location of the specific bone markings is facilitated if you first know the general names of the different types of markings.

BONE MARKINGS—GENERAL TYPES

Projections

1. **Trochanter**: A large irregularly shaped projection to which muscles attach.
2. **Tuberosity**: A large rounded projection to which muscles attach.
3. **Tubercle**: A small rounded projection to which muscles attach.
4. **Condyle**: A rounded convex projection.
5. **Head**: A rounded extension projecting from a tapering neck that forms a joint.
6. **Ramus**: An armlike branch extending from body of bone that forms a joint.
7. **Spine**: A sharp projection that serves as a point for muscle attachment.
8. **Crest**: An elevation or ridge serving as a point for muscle attachment.
9. **Line**: A lesser elevation or ridge that serves as a point of muscle attachment.
10. **Process**: Term used to designate a general outgrowth of varying shape and size.

Depressions

1. **Fossa**: A hollow indentation, often rounded, often found in joint formations.
2. **Foramen**: A hollow opening that serves as a passageway for nerves and blood vessels.
3. **Sinus**: Irregularly shaped space often filled with air and lined with mucosal tissue.
4. **Fissure**: A narrow slitlike depression often serving as a passageway for nerves.
5. **Meatus**: A canal-like opening that sometimes serves as a passageway for nerves.

The following table entitled "Identification of Bone Markings" includes the more generally known bone markings. Identify the bone markings. Follow the table sequentially. This systematized approach to study facilitates learning.

IDENTIFICATION OF BONE MARKINGS

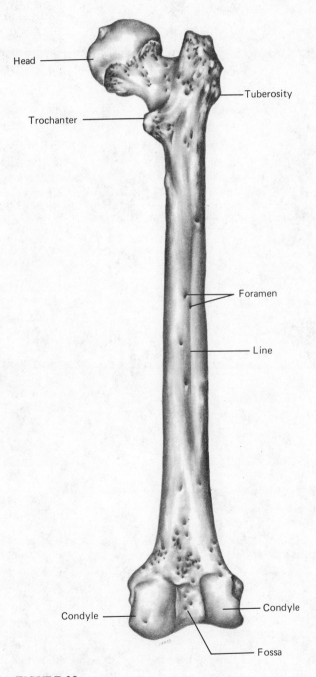

Head

Tuberosity

Trochanter

Foramen

Line

Condyle

Condyle

Fossa

FIGURE 30

Bone	Bone Marking	Description
A. Axial Skeleton		
Cranium		
Frontal		A flat convex bone in the adult, formed by the union of two flat bones joined at frontal suture (suture partly visible in adult)
(Anterior view— *Figure 31*)	Frontal eminences or protuberances	Protrusions above each optic orbit
	Supraorbital margins	Flat arched regions immediately inferior to eyebrows
	Supraorbital notches or foramina	Irregularly shaped small openings in supraorbital margins toward nasal bones, serving as an opening for nerves and blood vessels
	Superciliary arches	Anterior ridges superior to optic orbits that form base for eyebrows; ridges formed by frontal sinuses that lie within them
	Glabella	Smooth area formed by the union of the supercilliary arches above nasal bones
Parietal		Large irregularly shaped bones lateral and posterior to frontal bones; parietals merge at superior medial portion of the cranium to form sagittal suture
	Parietal foramen	Small rounded opening toward the posterior of the bone; immediately lateral to the sagittal suture
	Parietal tuberosity	Slightly rounded protrusion on lateral surface of bone
Temporal		Bones form lateral, inferior sides of cranium, houses structures of middle and inner ear
(Lateral view)	Mastoid process	Rough projection immediately posterior to ear or external auditory meatus
	Squamous portion	Flat, thin, superior portion of bone; transparent when held to light
	Zygomatic process	Thin, narrow, bridgelike extension that articulates anteriorly with zygomatic bone

Bone	Bone Marking	Description
	External auditory meatus	Canal into ear running from exterior to interior of temporal bone
	Mandibular fossa	Rounded depression inferior to and partially formed by zygomatic process; immediately anterior to external auditory meatus; forms socket for mandibular condyle

FIGURE 31 *Anterior aspect of human skull*

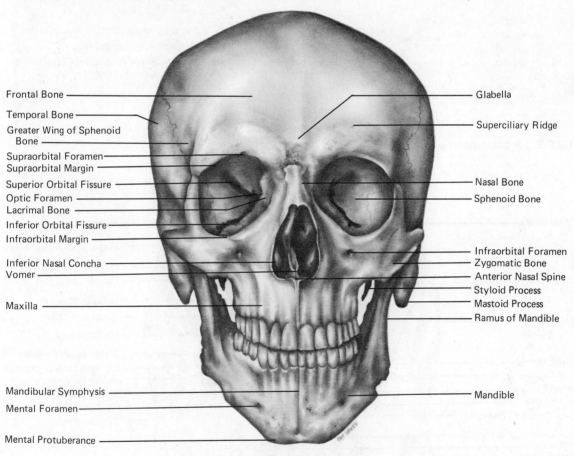

Frontal Bone

Temporal Bone

Greater Wing of Sphenoid Bone

Supraorbital Foramen

Supraorbital Margin

Superior Orbital Fissure

Optic Foramen

Lacrimal Bone

Inferior Orbital Fissure

Infraorbital Margin

Inferior Nasal Concha

Vomer

Maxilla

Mandibular Symphysis

Mental Foramen

Mental Protuberance

Glabella

Superciliary Ridge

Nasal Bone

Sphenoid Bone

Infraorbital Foramen

Zygomatic Bone

Anterior Nasal Spine

Styloid Process

Mastoid Process

Ramus of Mandible

Mandible

Bone	Bone Marking	Description
(Inferior view— *Figure 32*) Skull viewed from below	Styloid process	Long needlelike projection, anterior to mastoid process; serves as attachment for muscles and ligaments
	Stylomastoid foramen	Small opening between styloid process and mastoid process; serves as a passageway for nerve
	Jugular fossa	Irregularly shaped depression lateral and anterior to occipital condyles; opening for internal jugular vein
	Jugular foramen	Fairly large oval-shaped opening at posterior end of jugular fossa; lateral to occipital condyle; serves as passageway for lateral sinus and IX, X, and XI cranial nerves
	Carotid foramen or canal	Round openings medial and anterior to jugular foramen; serves as passageway for internal carotid artery

FIGURE 32 *Inferior aspect of human skull*

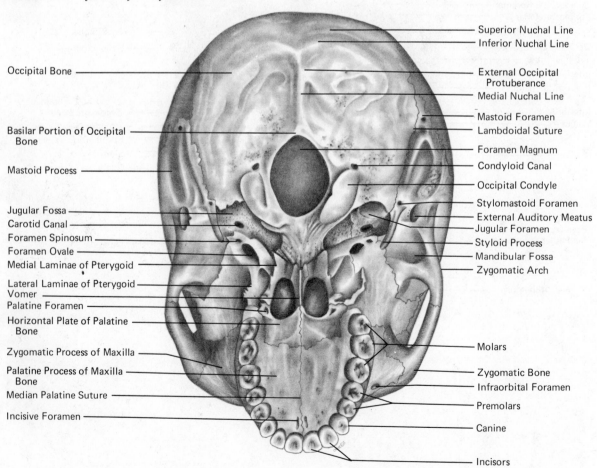

Bone	Bone Marking	Description
(Internal view—*Figure 33*) View of floor of cranial cavity; superior section of cranium removed	Petrous portion	Rock-like protrusion in center of cranial floor; flares from medial to lateral region; middle and inner ear structures are contained within
	Internal auditory meatus	Opening in medial side of petrous portion, it is the internal opening of canal running from external auditory meatus through which the acoustic nerve passes
Occipital		Large flaring bone forming inferior, posterior section of cranium; articulates with parietal bones to form lambdoidal suture
(Inferior view—*Figure 32*)	External occipital protuberance	Protrusion or eminence that extends most posteriorly in medial portion: bump-like
	Superior nuchal line	Elevation projecting laterally from external occipital protuberance

FIGURE 33 *Interior aspect of human skull*

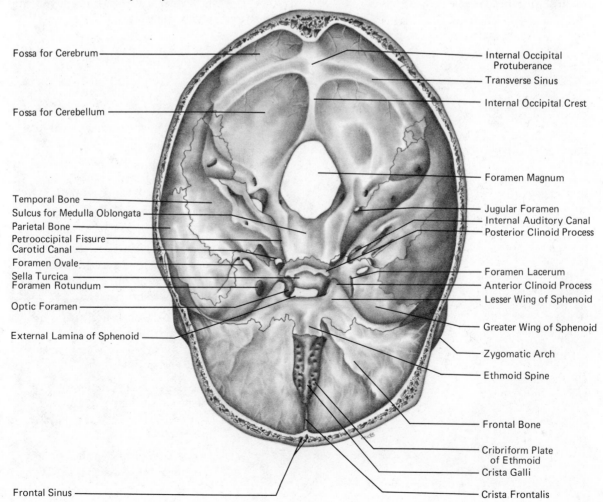

Fossa for Cerebrum

Fossa for Cerebellum

Temporal Bone
Sulcus for Medulla Oblongata
Parietal Bone
Petrooccipital Fissure
Carotid Canal
Foramen Ovale
Sella Turcica
Foramen Rotundum
Optic Foramen

External Lamina of Sphenoid

Frontal Sinus

Internal Occipital Protuberance
Transverse Sinus
Internal Occipital Crest

Foramen Magnum

Jugular Foramen
Internal Auditory Canal
Posterior Clinoid Process

Foramen Lacerum
Anterior Clinoid Process
Lesser Wing of Sphenoid

Greater Wing of Sphenoid

Zygomatic Arch

Ethmoid Spine

Frontal Bone

Cribriform Plate of Ethmoid
Crista Galli

Crista Frontalis

Bone	Bone Marking	Description
	Inferior nuchal line	A lesser elevation inferior to superior nuchal line
	Condyles	Oval-shaped protrusions with flat surfaces lateral to foramen magnum; articulate with atlas
	Foramen magnum	Large round opening that serves as passageway for spinal cord in its connection with brain; medial to condyles
(Internal view— *Figure 33*) View of floor of cranial cavity	Internal occipital protuberance	Linear projection in internal medial region of bone
Sphenoid		Bat-shaped bone forming part of the floor of the cranial cavity; seen laterally anterior to the temporal bone
	Body	Medial cubelike region; hollow centrally
(Internal view— *Figure 33*) View of floor of cranial cavity	Greater wings	Laterally flaring portion forming floor of cranial cavity and outer wall of optic orbit
	Lesser wings	Thin flaring portions superior to greater wings and body; form superior posterior section of optic orbit
	Sella turcica	Saddle-shaped indention medial to wings; houses pituitary gland
	Optic foramen	Rounded opening inferior to lesser wings at medial point; can be seen anteriorly through optic orbit; transmits II cranial nerve
	Superior orbital fissure	Irregularly shaped furrow, immediately inferior and covered by lesser wings; can be seen anteriorly through optic orbit; transmits III, IV, and V cranial nerves
	Foramen rotundum	Small rounded hole in greater wings; lateral to sella turcica; transmits maxillary division of V (trigeminal) cranial nerve
	Foramen ovale	Oval-shaped hole posterior to foramen rotundum; transmits mandibular division of trigeminal nerve
	Foramen lacerum	Oval-shaped opening in greater wing toward body; lateral to sella

Bone	Bone Marking	Description
		turcica at apex of petrous portion; transmits internal carotid arteries and branch of ascending pharyngeal artery
	Foramen spinosum	Small irregularly shaped opening posterior and lateral to foramen ovale; transmits mandibular nerves and meningeal arteries
(Lateral view)	Pterygoid process	Wing-shaped downward projection posterior to maxilla and molar teeth; consists of medial and lateral lamina
Ethmoid		Irregularly shaped bone that forms anterior superior section of cranial floor; medial posterior walls of optic orbits; lateral walls, posterior septum and roof of nasal cavity
	Crista galli	Flaglike structure pointing anteriorly; serves as a point of attachment for meninges
	Horizontal plate (Cribriform plate)	Porous flat section of bone lateral to crista galli; transmits olfactory nerves
	Perpendicular plate	Linear projection extending inferiorly from crista galli
	Lateral masses	Lateral portions of ethmoid; inner section forms walls of nasal cavities and conchae; contains many air cavities
	Ethmoid sinus	Air cavity within lateral masses
	Superior and middle conchae (Turbinates)	Extensions from lateral sides of nasal cavity pointing medially toward septum; contain spongy air-filled cavities
Face Bones		
Mandible		Lower jaw; articulates with temporal bone; forms the only diarthrotic joint in cranium
	Body	Rounded anterior section; convex posteriorly; forms chin
	Mental tubercles	Small rounded protrusions on inferior rim of body; on external surface
	Mental foramen	Small rounded opening on external surface below bicuspids; transmits nerves and blood vessels
	Angle	Junction of body and ramus on inferior side

Bone	Bone Marking	Description
	Ramus	Superior armlike projections on either side of body; posterior part of bone
	Mandibular foramen	Opening on internal surface of ramus; passageway for nerves and blood vessels to teeth of mandible
	Mandibular condyle (Condyloid process)	Posterior rounded head that fits into mandibular fossa of temporal bone
	Mandibular notch	Depression between condyloid process and coronoid process
	Neck	Constricted projection at base of condyloid process
	Coronoid process	Thin, pointed anterior projection that serves as point for muscle attachment
	Alveolar process	Sockets; process of bone into which teeth are anchored centrally
Nasal (frontal view)		Small irregularly shaped bones inferior to glabella; form upper division of nasal bridge
Lacrimal		Thin delicate bones lateral to nasal bones; forms medial wall of optic orbit; forms lateral wall of nasal cavity
	Groove	Slight indentation or furrow running in a superior to inferior direction; serves as a passageway for tears
Vomer		Single irregularly shaped bone found medially and posteriorly in the nasal cavity
Inferior conchae (turbinates)		Thin delicate bones flaring from lateral walls of nasal cavity; most posterior projections
Zygomatic (malar)		Hard, heavy bone; cheekbone
	Zygomaticofacial foramen	Small round opening anterior to zygomatic arch; inferior and slightly lateral to optic orbit
Palatine		L-shaped bone that forms the posterior part of the hard palate; also forms lateral and inferior walls of posterior division of nasal cavity
	Horizontal plate	Horizontal junction with maxilla; forms most posterior section of hard palate

Bone	*Bone Marking*	*Description*
Maxilla		Upper jaw bone, main bone of face; all other facial bones articulate with maxilla
(Posterior view)	Antrum of Highmore (Maxillary sinus)	Largest of all sinus, in central cavity of maxilla; mucus lined and air filled
	Incisive foramen	Triangular opening between central incisors; located at midline of maxilla
	Palatine process	Forms most of hard palate; curves downward to form alveoli
(Anterior view)	Alveolar processes	Sockets containing teeth; archlike
	Anterior nasal spine	Projection inferior and anterior to nasal septum
Hyoid		U-shaped bone anteriorly located between larynx and mandible; serves as point of muscle attachment for some muscles of mouth and tongue
	Transverse body	Rounded anterior portion of bone
	Greater cornu	Posterior projections extending from transverse body
	Lesser cornu	Anterior projections on either side of the anterior portion of the transverse body
Vertebrae		Twenty-six blocklike bones comprising the vertebral column, which protects the spinal cord and allows for anterior, posterior, and lateral movement
Typical Structures of All Vertebrae		
Body		Rounded central bony portion facing anteriorly in column; points of attachment for intervertebral disks
Vertebral arch		Junction of all posterior extensions from body
Vertebral foramen		Central opening through which spinal cord passes
Pedicle		Thick, bony, lateral extension posterior to body; forms lateral portion of vertebral foramen
Lamina		Flat, thinner, bony extension; extends pedicle toward midline posteriorly; forms posterior portion of vertebral foramen

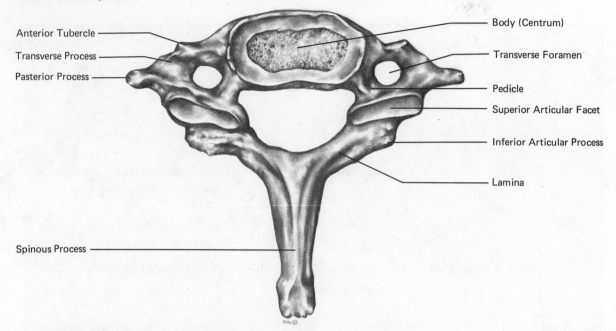

Anterior Tubercle

Transverse Process

Pasterior Process

Body (Centrum)

Transverse Foramen

Pedicle

Superior Articular Facet

Inferior Articular Process

Lamina

Spinous Process

FIGURE 34 *Superior aspect of typical cervical vertebra*

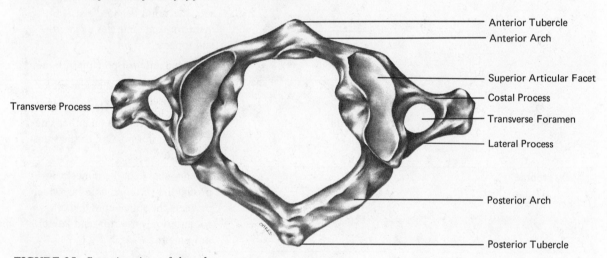

Anterior Tubercle

Anterior Arch

Superior Articular Facet

Costal Process

Transverse Foramen

Lateral Process

Posterior Arch

Posterior Tubercle

Transverse Process

FIGURE 35 *Superior view of the atlas*

FIGURE 36 *Inferior aspect of atlas*

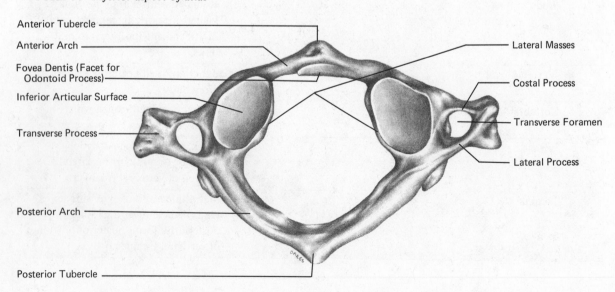

Anterior Tubercle

Anterior Arch

Fovea Dentis (Facet for Odontoid Process)

Inferior Articular Surface

Transverse Process

Posterior Arch

Posterior Tubercle

Lateral Masses

Costal Process

Transverse Foramen

Lateral Process

Bone	Bone Marking	Description
Intervertebral foramen		Opening between vertebrae for emergence of spinal nerves
Processes		Outgrowths of bone
Transverse processes		Lateral extensions from junction of pedicle and lamina
Superior articulating surfaces		Paired projections on superior surface lateral to vertebral foramen; slightly indented centrally; form articulation with immediately superior vertebra (or, in case of atlas, articulated with occipital condyles)
Inferior articulating surfaces		Paired projections on inferior surface lateral to vertebral foramen; slightly indented centrally; form articulation with immediately inferior vertebra
Spinous process		Single posterior and inferiorly extending process at midline of vertebra
Specific Vertebral Structures		
Cervical (*Figure 34*)		All of the typical structures; smallest lightest vertebrae; slightly larger, triangular, vertebral foramen; spinous processes bifurcated and short; transverse processes contain foramen (transverse foramen)
Atlas (First cervical vertebra) (*Figures 35* and *36*)		No body; lateral processes contain large concave articulating surfaces

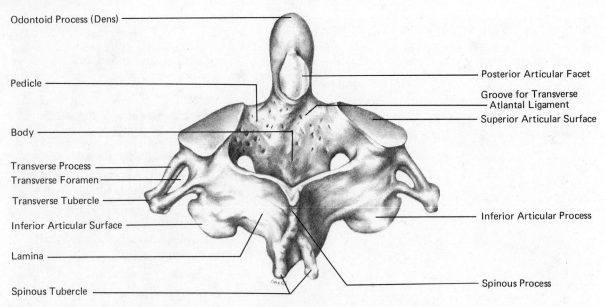

Odontoid Process (Dens)

Pedicle

Body

Transverse Process
Transverse Foramen

Transverse Tubercle

Inferior Articular Surface

Lamina

Spinous Tubercle

Posterior Articular Facet

Groove for Transverse
Atlantal Ligament

Superior Articular Surface

Inferior Articular Process

Spinous Process

FIGURE 37 *Dorsal aspect of axis*

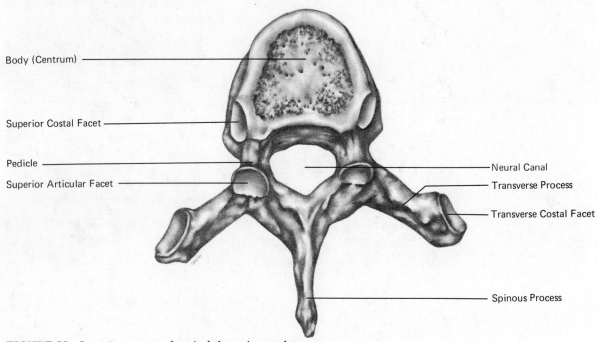

Body (Centrum)

Superior Costal Facet

Pedicle

Superior Articular Facet

Neural Canal

Transverse Process

Transverse Costal Facet

Spinous Process

FIGURE 38 *Superior aspect of typical thoracic vertebra*

FIGURE 39 *Lateral aspect of 12th thoracic vertebra*

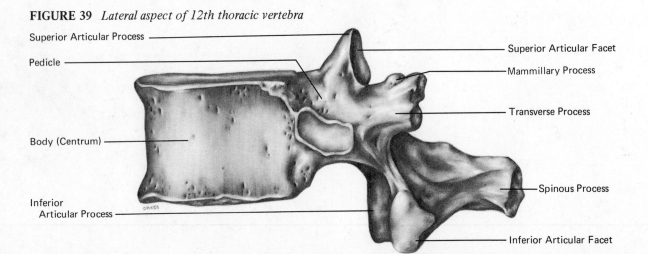

Superior Articular Process

Pedicle

Body (Centrum)

Inferior
Articular Process

Superior Articular Facet

Mammillary Process

Transverse Process

Spinous Process

Inferior Articular Facet

Bone	Bone Marking	Description
Axis (Second cervical vertebra) (*Figure 37*)		Acts as a pivot for rotation of atlas and cranium; *odontoid process* is a superior vertical projection serving as pivot point and body
Thoracic (*Figures 38–39*)		All of the typical structures; larger body than cervical; oval vertebral foramen; sharp inferiorly projecting spinous process; contains lateral articulating surfaces, superior and inferior for rib articulation
Lumbar (*Figure 40*)		All of the typical structures; large, thick, blocklike body with superior articulating surfaces projecting inward and inferior articulating surfaces outward; short, thick, spinous process projecting posteriorly
Sacrum (*Figure 50*)		Fusion of five vertebrae; large, flat, and slightly concave bone forming posterior border of pelvic brim
Coccyx (*Figure 50*)		Fusion of three to five small irregularly shaped vertebrae forming taillike projection inferiorly

FIGURE 40 *Superior aspect of typical lumbar vertebra*

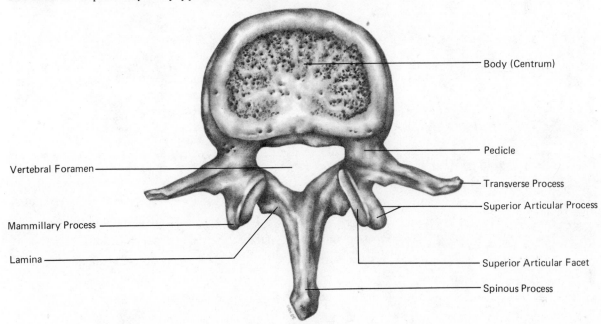

Body (Centrum)

Pedicle

Vertebral Foramen

Transverse Process

Superior Articular Process

Mammillary Process

Lamina

Superior Articular Facet

Spinous Process

Bone	Bone Marking	Description
Sternum		Breastbone; fusion of three bones; T-shaped bone articulating with true ribs
Manubrium		Superior horizontal portion of bone
Gladiolus		Central vertical portion of bone
Xiphoid process		Cartilaginous inferior projection; pointed
Ribs (seven pairs true, five pairs false)		Form walls of thoracic cavity; true ribs articulate with both vertebrae and sternum; false ribs articulate with vertebrae and have cartilaginous connections to sternum, except for last two pairs of "false ribs" which are floating ribs and have no anterior connection to sternum
	Head	Posterior rounded projection, articulating with a thoracic vertebra
	Neck	Narrow portion below head and connecting head with shaft
	Shaft (body)	Long, thin, curved portion
	Tubercle	Small rounded protrusion below neck on shaft
	Costal cartilage	Cartilaginous connection of ribs to sternum
B. Appendicular Skeleton		
Upper Division		
Clavicle		Collarbone; S-shaped body articulating posteriorly with scapula and anteriorly with sternum
	Sternal end	Rounded end that articulates with sternum
	Acromial end	Flat end that articulates with acromion process of scapula
Scapula (*Figures 41–42*)		Two wing-shaped bones on either side of body midline posteriorly; shoulder bones
	Body	Flat, thin central portion
	Superior border	Upper margin of body
	Vertebral border	Medial margin continuous with vertebral column
	Axillary border	Lateral margin

Bone	Bone Marking	Description
	Spine	Posterior ridgelike projection extending from vertebrae toward axillary border
	Acromial process or acromion	Articulation with clavicle; rounded end of spine
	Coracoid process	Lateral rounded projection on superior border
	Glenoid cavity	Articulation point for humerus; rounded fossa on superior end of axillary border

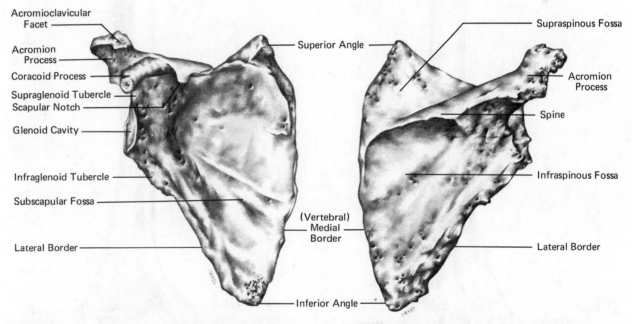

FIGURE 41 *Anterior view of right scapula*

FIGURE 42 *Posterior view of right scapula*

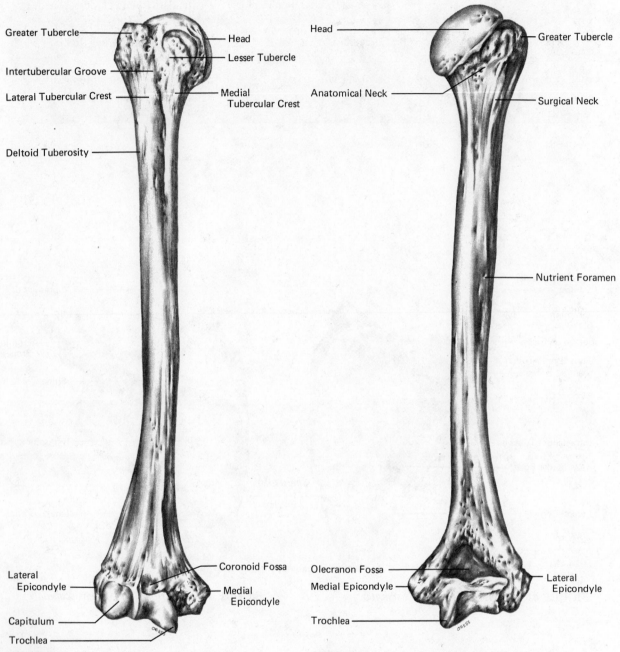

Greater Tubercle

Head

Lesser Tubercle

Intertubercular Groove

Lateral Tubercular Crest

Medial Tubercular Crest

Deltoid Tuberosity

Head

Greater Tubercle

Anatomical Neck

Surgical Neck

Nutrient Foramen

Lateral Epicondyle

Coronoid Fossa

Medial Epicondyle

Capitulum

Trochlea

Olecranon Fossa

Medial Epicondyle

Lateral Epicondyle

Trochlea

FIGURE 43 *Anterior aspect of right humerus*

FIGURE 44 *Posterior aspect of right humerus*

Bone	Bone Marking	Description
Humerus (*Figures 43–44*)		Upper arm bone articulating with scapula proximally and with ulna and radius distally
	Head	Rounded posteriomedial projection on proximal end
	Anatomical neck	Oblique constriction on anterior surface below head
	Greater tubercle	Larger anteriolateral projection from head
	Lesser tubercle	Smaller anterior projection below anatomical neck
	Intertubercular groove	Indention between greater and lesser tubercles
	Surgical neck	Posterior constriction below tubercles; often fractured
	Deltoid tuberosity	Rough lateral protrusion on mid-diaphysis
	Medial epicondyle	Rough rounded projection on medial surface at distal end
	Lateral epicondyle	Rough smaller protrusion on lateral surface at distal end
	Capitulum (radial head)	Lateral rounded knob immediately anterior to lateral epicondyle that articulates with radius
	Trochlea	Pulley-shaped projection medial and inferior to medial epicondyle that articulates with ulna
	Olecranon fossa	Deep triangular depression on posterior surface between epicondyles; articulates with olecranon process during extension of forearm
	Coronoid fossa	Depression on anterior surface immediately superior to trochlea; articulates with coronoid process of forearm during flexion

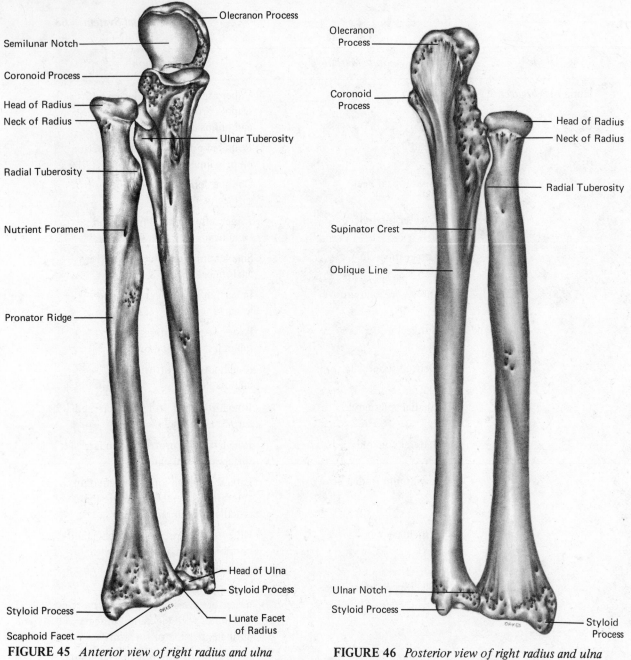

FIGURE 45 *Anterior view of right radius and ulna*

Labels (Figure 45):
- Olecranon Process
- Semilunar Notch
- Coronoid Process
- Head of Radius
- Neck of Radius
- Radial Tuberosity
- Nutrient Foramen
- Pronator Ridge
- Ulnar Tuberosity
- Head of Ulna
- Styloid Process
- Lunate Facet of Radius
- Styloid Process
- Scaphoid Facet

FIGURE 46 *Posterior view of right radius and ulna*

Labels (Figure 46):
- Olecranon Process
- Coronoid Process
- Head of Radius
- Neck of Radius
- Radial Tuberosity
- Supinator Crest
- Oblique Line
- Ulnar Notch
- Styloid Process
- Styloid Process

Bone	Bone Marking	Description
Ulna (*Figures 45–46*)		Medial bone of forearm; longer than radius
	Olecranon process	Large round projection at proximal end; center concave (semilunar notch) resembles "open mouth"; forms elbow
	Coronoid process	Protrusion on anterior surface below olecranon process; lower half of "open mouth"; articulates with humerus

64

Bone	Bone Marking	Description
	Semilunar notch	Concave notch that articulates with trochlea of humerus
	Radial notch	Concave indention that articulates with lateral aspect of radial head
	Styloid process	Small V-shaped projection at distal end of bone
Radius (*Figures 45–46*)		Shorter, lateral bone of forearm
	Radial head	Rounded knoblike projection at proximal end of bone
	Radial tuberosity	A raised rough protrusion on the medial aspect near the proximal end
	Styloid process	A needlelike projection on the lateral aspect of the distal end
Carpals (*Figure 47*)		Sixteen irregularly shaped bones; eight comprising each wrist

FIGURE 47 *Dorsal aspect of bones of right hand*

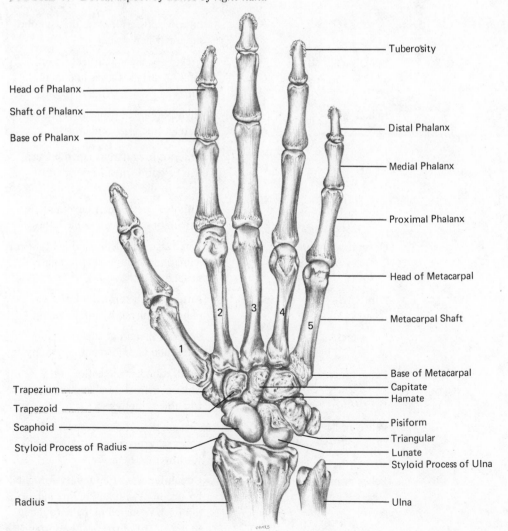

Bone	Bone Marking	Description
Lower Division		
Innominate pelvic bones (*Figures 48–49*)	(Ossa coxae)	Fusion of three smaller bones to form right and left pelvic bones
Ilium		Superior flaring portion of bone
	Iliac crests	Superior border of ilium
	Iliac spines	Projections on periphery of ilium (named by position)
	Anterior superior	Most anterior projection of hip
	Anterior inferior	Less prominent projection inferior to anterior superior spine
	Posterior superior	Most posterior projection on periphery of iliac crest
	Posterior inferior	Less prominent projection inferior to posterior superior spine
	Greater sciatic notch	Large irregularly shaped indentation inferior to posterior inferior spine
	Iliac fossa	Concave inner surface of ilium
Ischium		Inferior, posterior bone that supplies support when sitting
	Ischial tuberosity	Irregularly shaped projection on inferior surface that is in contact with chair when sitting
	Ischial spine	Slightly pointed projection immediately superior to tuberosity
Pubis		Anterior irregularly shaped bones; two fuse to form anterior part of pelvic girdle
	Pubic symphysis	Anterior, mesial, cartilaginous articulation between pubic bones
	Obturator foramen	Very large rounded opening between pubis and ischium on anteriolateral surface
	Pubic arch	Anterior arch formed inferior to pubic symphysis by inferior rami
	Pubic crest	Superior margin of superior rami above and lateral to pubic symphysis
	Pubic tubercle	Small rounded projections on superior rami; lateral and superior to pubic symphysis
	Pelvic girdle	Articulation of ossa coxa, sacrum, and coccyx; forms point of attachment for lower extremities
	Pelvic brim	Circumference of boundary formed by fusion of ilium, ischium, and pubis; oval shaped in male, rounded in female; of obstetrical importance in childbirth

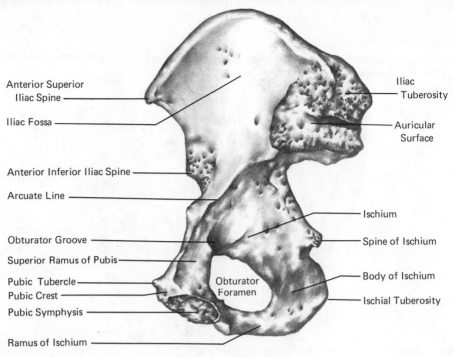

FIGURE 48 *Medial aspect of right hipbone*

Anterior Superior Iliac Spine

Iliac Fossa

Anterior Inferior Iliac Spine

Arcuate Line

Obturator Groove

Superior Ramus of Pubis

Pubic Tubercle

Pubic Crest

Pubic Symphysis

Ramus of Ischium

Iliac Tuberosity

Auricular Surface

Ischium

Spine of Ischium

Body of Ischium

Ischial Tuberosity

Obturator Foramen

FIGURE 49 *Lateral aspect of right hipbone*

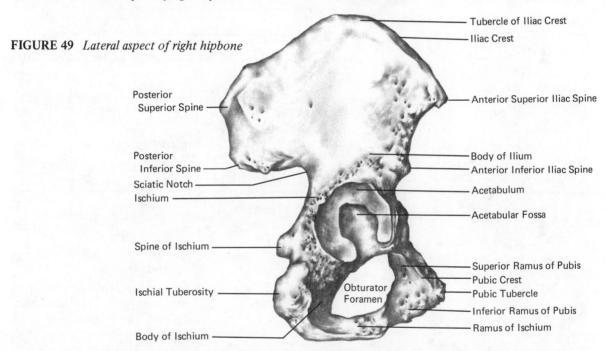

Tubercle of Iliac Crest

Iliac Crest

Posterior Superior Spine

Posterior Inferior Spine

Sciatic Notch

Ischium

Spine of Ischium

Ischial Tuberosity

Body of Ischium

Anterior Superior Iliac Spine

Body of Ilium

Anterior Inferior Iliac Spine

Acetabulum

Acetabular Fossa

Superior Ramus of Pubis

Pubic Crest

Pubic Tubercle

Inferior Ramus of Pubis

Ramus of Ischium

Obturator Foramen

Bone	Bone Marking	Description
	True pelvis	Space inferior to pelvic brim; contains pelvic organs
	False pelvis	Space superior to pelvic brim; medial to iliac bones; included in abdominal cavity
	Acetabulum	Lateral cavity formed by fusion of ilium, ischium, and pubis; hip socket, articulates with femur head
Sacrum (*Figure 50*)		A fusion in the adult of the five sacral vertebrae which with the coccyx form the dorsal portion of the pelvic girdle.

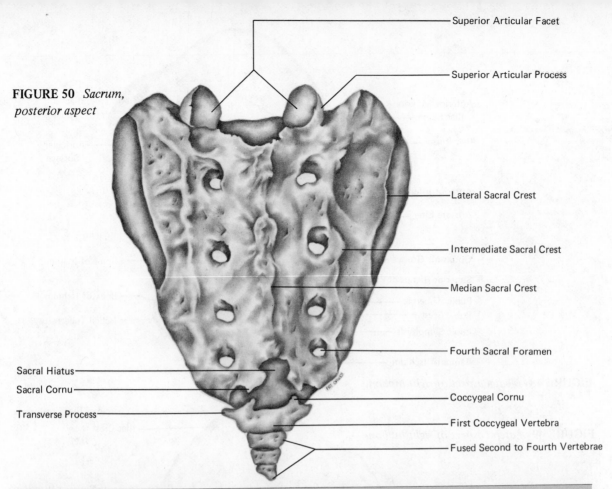

FIGURE 50 *Sacrum, posterior aspect*

Superior Articular Facet

Superior Articular Process

Lateral Sacral Crest

Intermediate Sacral Crest

Median Sacral Crest

Fourth Sacral Foramen

Sacral Hiatus

Sacral Cornu

Transverse Process

Coccygeal Cornu

First Coccygeal Vertebra

Fused Second to Fourth Vertebrae

Bone	Bone Marking	Description
Femur (*Figures 51–52*)		Large strong thigh bone
	Head	Rounded projection that articulates with acetabulum in ball and socket joint
	Greater trochanter	Large irregularly shaped process inferior and lateral to head
	Lesser trochanter	Small irregularly shaped process inferior and posterior to head
	Linea aspera	Ridge running approximately three-fourths the length of posterior bone surface
	Supracondylar ridges	Inferior division of linea aspera into two ridges
	Gluteal tubercle	Small rounded projection inferior to greater trochanter
	Medial condyle	Large rounded protrusion on medial distal end
	Lateral condyle	Large rounded protrusion on lateral distal end
	Adductor tubercle	Very small rough projection above medial condyle marking termination of medial supracondylar ridge
	Trochlea	Depression on anterior surface between condyles
	Intercondyloid notch	Deep depression between condyles on posterior surface
Patella		Irregularly shaped flat knee bone

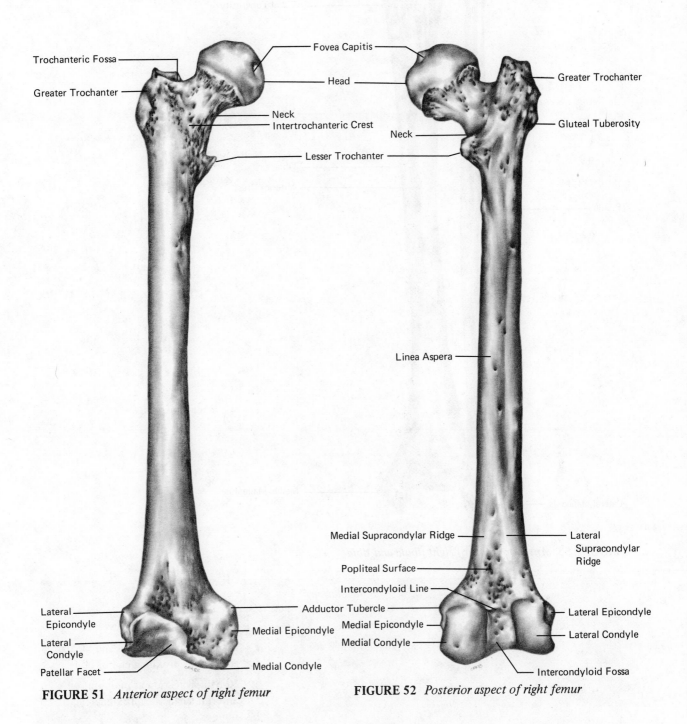

Trochanteric Fossa

Greater Trochanter

Fovea Capitis

Head

Neck

Intertrochanteric Crest

Lesser Trochanter

Greater Trochanter

Gluteal Tuberosity

Neck

Linea Aspera

Medial Supracondylar Ridge

Popliteal Surface

Intercondyloid Line

Adductor Tubercle

Medial Epicondyle

Medial Condyle

Lateral Epicondyle

Lateral Condyle

Patellar Facet

Medial Epicondyle

Medial Condyle

Lateral Supracondylar Ridge

Lateral Epicondyle

Lateral Condyle

Intercondyloid Fossa

FIGURE 51 *Anterior aspect of right femur*

FIGURE 52 *Posterior aspect of right femur*

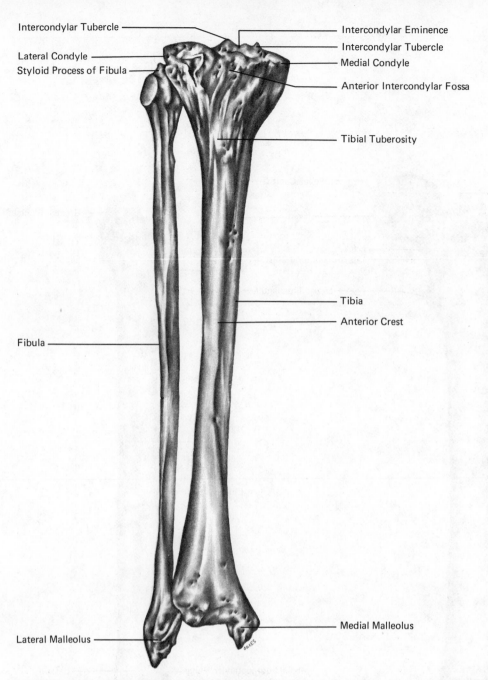

Intercondylar Tubercle

Lateral Condyle

Styloid Process of Fibula

Intercondylar Eminence

Intercondylar Tubercle

Medial Condyle

Anterior Intercondylar Fossa

Tibial Tuberosity

Tibia

Anterior Crest

Fibula

Medial Malleolus

Lateral Malleolus

FIGURE 53 *Anterior aspect of right fibula and tibia*

Bone	Bone Marking	Description
Tibia (*Figure 53*)		Large strong bone of lower leg
	Lateral and medial condyles	Protuberance on lateral and medial surfaces at proximal end of bone
	Tibial tuberosity	Rough projection at midpoint on anterior surface at proximal end of bone
	Intercondylar eminence	Toothlike projection on superior surface between condyles
	Popliteal line	Slight ridge on posterior surface running from superior to inferior; extends approximately one-third the length of the bone
	Medial malleolus	Medial protrusion on distal end

Bone	Bone Marking	Description
Fibula (*Figure 53*)		Long thin bone of lower leg
	Lateral malleolus	Rounded projection on lateral surface of bone
Tarsals		Fourteen irregularly shaped bones of upper foot—seven per foot
Calcaneus		Large irregular bone that forms heel
Talus (*Figure 54*)		Rounded bone with smooth superior surface that articulates with tibia and fibula
Metatarsals		Ten long bones articulating between the tarsals and proximal phalanges—five per foot

FIGURE 54 *Bones of right foot, dorsal surface*

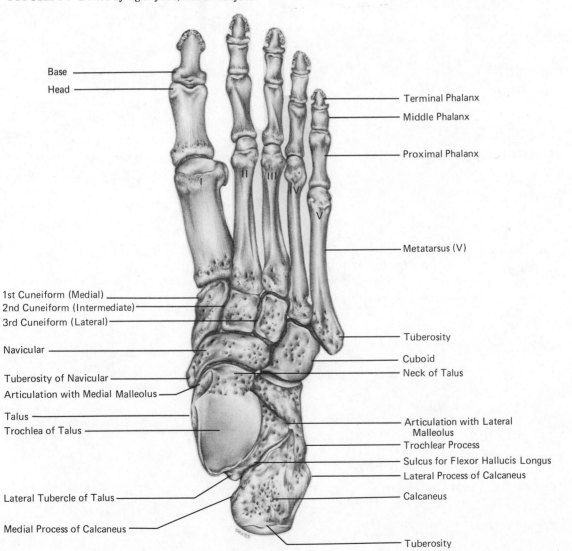

Skeletal System

DISCUSSION

1. List the structures of a Haversian system and give the function of each.

2. Define the following bone markings and give an example of each.

a. trochlea _____

_____

b. tubercle _____

_____

c. trochanter _____

_____

d. spine _____

_____

e. process _____

_____

f. fossa _____

_____

g. foramen _____

_____

h. condyle _____

_____

i. groove _____

_____

3. Which bones are included in (a) the axial skeleton and (b) the appendicular skeleton?

4. List several reasons stating the importance of knowing the major bone markings.

5. Give the function of the following bone markings:

   a. mandibular fossa _____

   b. jugular foramen _____

   c. carotid foramen _____

   d. sella turcica _____

   e. superior orbital fissure _____

   f. foramen rotundum _____

   g. crista galli _____

   h. foramen magnum _____

   i. superior articulating surface of a vertebra _____

   j. odontoid process _____

   k. rib tubercle _____

   l. acromion of scapula _____

   m. glenoid cavity _____

   n. coronoid fossa of humerus _____

   o. semilunar notch of ulna _____

   p. ischial tuberosity _____

   q. acetabulum _____

   r. trochlea of femur _____

   s. intercondylar eminence _____

   t. trochlea of humerus _____

   u. sternal end of clavicle _____

   v. costal cartilage of rib _____

UNIT VI *Muscular System*

A. Fetal Pig Musculature

PURPOSE
The purpose of Unit VI-A is to enable the student to understand the gross and microscopic anatomical relationships of muscles using the fetal pig as a specimen for dissection.

OBJECTIVES
In order to complete Unit VI-A, the student must be able to do the following:
1. Properly remove the skin, underlying fat, and fascia from the fetal pig.
2. Identify selected muscles of the head, trunk, and limbs.
3. Microscopically recognize skeletal, smooth, and cardiac muscle.
4. Microscopically recognize a myoneural junction.
5. Recognize similarities and differences in muscles found in the fetal pig and human being.

MATERIALS

preserved fetal pigs
forceps
sharp scissors
dissecting pins
newspaper
slides of myoneural junction

sharp scalpel, preferably
   with removable blade

slides of myoneural
   junction

slides of skeletal, smooth,
   and cardiac muscle

PROCEDURE

EXERCISE 1 *Skinning the Fetal Pig*

Your instructor will demonstrate the proper way to skin your specimen, remove fat, fascia (connective tissue), and to separate selected muscles. The muscles of the fetal pig are not fully developed and are easily torn. Therefore, care must be exercised in skinning and dissecting this animal. The usual procedure for skinning a fetal pig is to make a midventral incision at the base of the throat with scissors, and being sure you are cutting only through the skin, extend the incision toward the hind legs, cutting around the umbilicus. Then extend your initial midventral incision down the medial surface of each leg and cut around the wrists and ankles. Using forceps or your fingers, grasp the cut edge of the skin and gently pull it away from the body, using the handle of your scalpel or your other hand to separate the skin from underlying connective tissue and muscles. The midventral incision may also be extended in the other direction, exposing the muscles of the neck and throat. Although you will have skinned the entire specimen, it is necessary to identify muscles on one side only. The other side may be used for reviewing the musculature or locating blood vessels in a later unit.

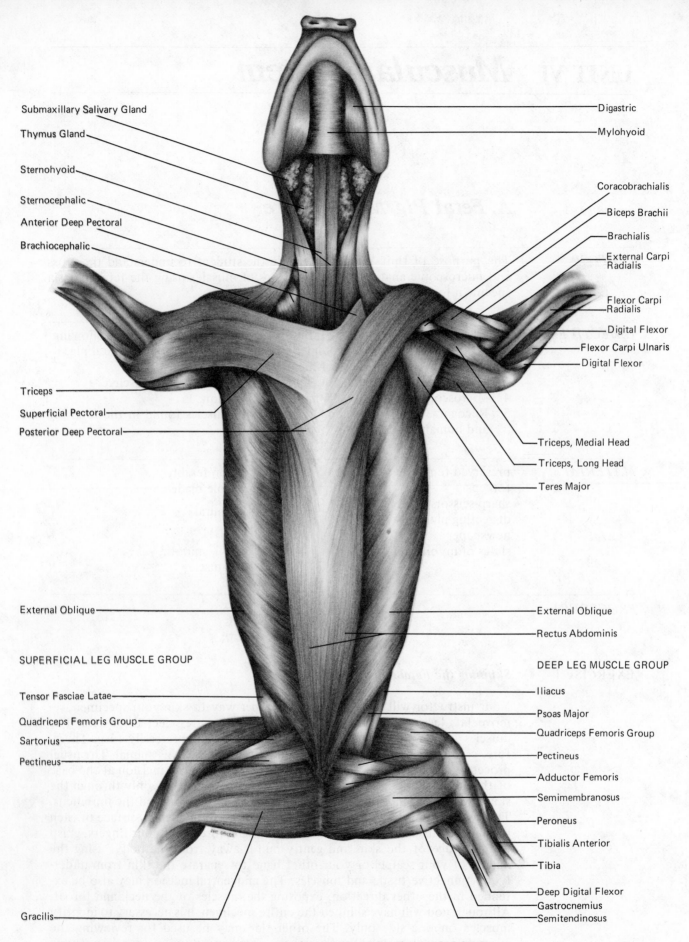

Submaxillary Salivary Gland

Thymus Gland

Sternohyoid

Sternocephalic

Anterior Deep Pectoral

Brachiocephalic

Triceps

Superficial Pectoral

Posterior Deep Pectoral

External Oblique

SUPERFICIAL LEG MUSCLE GROUP

Tensor Fasciae Latae

Quadriceps Femoris Group

Sartorius

Pectineus

Gracilis

Digastric

Mylohyoid

Coracobrachialis

Biceps Brachii

Brachialis

External Carpi Radialis

Flexor Carpi Radialis

Digital Flexor

Flexor Carpi Ulnaris

Digital Flexor

Triceps, Medial Head

Triceps, Long Head

Teres Major

External Oblique

Rectus Abdominis

DEEP LEG MUSCLE GROUP

Iliacus

Psoas Major

Quadriceps Femoris Group

Pectineus

Adductor Femoris

Semimembranosus

Peroneus

Tibialis Anterior

Tibia

Deep Digital Flexor

Gastrocnemius

Semitendinosus

FIGURE 55 *Ventral aspect of fetal pig, superficial muscles of the chest and abdomen*

EXERCISE 2   *Identification of Muscles*

You will be responsible for the separation and identification of muscles in the major body regions. Knowledge of the general description, origin, insertion, and action of each muscle will help in its identification. It is necessary that fat and excess fascia be removed and each muscle be separated so it can be readily identified.

MUSCLES OF THE ABDOMINAL WALL (See Figure 55)

Name and Description	Origin	Insertion	Action
External Oblique: A superficial sheet of muscle forming the outermost muscle of the abdominal wall	Lumbodorsal fascia and lateral surfaces of ribs	Aponeurosis into linea alba	Constricts abdomen
Internal Oblique: A sheetlike muscle immediately beneath the external oblique; its fibers run in a different direction from those of the external oblique	Lumbodorsal fascia	Aponeurosis into linea alba	Compressor of abdomen
Transversus Abdominis: The sheet of muscle beneath the internal oblique; fibers run in a different direction than those of the internal oblique	Lumbodorsal fascia	Aponeurosis into linea alba	Compressor of abdomen
Rectus Abdominis: A superficial longitudinal strip of muscle lying immediately lateral to the mid-ventral line of the abdomen	Pubic symphysis	Sternum and costal cartilages	Constricts abdomen

MUSCLES OF THE NECK AND THROAT (See Figure 55)

Name and Description	Origin	Insertion	Action
Sternocephalic: Also known as the Sterno-mastoid muscle. A narrow muscle extending from the dorsal aspect of the skull to the sternum	Sternum	Mastoid process	Singly, turns the head; together, lowers head

Name and Description	*Origin*	*Insertion*	*Action*
Sternothyrohyoid: A long, slender muscle on either side of the midline of the throat	Sternum	Laryngeal prominence and hyoid bone	Retracts and depresses the hyoid bone, base of tongue, and larynx
Masseter: Cheek muscle	Zygomatic arch	Lateral surface of mandible	Elevates mandible

MUSCLES OF THE CHEST (See Figure 55)

Name and Description	*Origin*	*Insertion*	*Action*
Superficial Pectoral: A flat band of muscle extending laterally from the sternum to the humerus	Ventral sternum	Humerus	Adducts forelimb
Posterior Deep Pectoral: A diagonal band of muscle deep to the superficial Pectoral muscle	Ventral sternum and costal cartilages 4 through 9	Proximal humerus	Adducts and retracts humerus
Anterior Deep Pectoral: A thin strip of muscle extending from immediately superior to the posterior deep pectoral, over the shoulder to the dorsal scapular region	Anterior sternum	Aponeurosis of dorsal portion of spinatus	Adducts and retracts forelimb

MUSCLES OF THE UPPER BACK, SHOULDER, AND BACK OF NECK (See Figure 56)

Name and Description	*Origin*	*Insertion*	*Action*
Latissimus Dorsi: A flat superficial muscle extending from the mid-dorsal line to the medial surface of the humerus	Thoracic and lumbar vertebrae; also lumbodorsal fascia	Medial surface of proximal end of humerus	Draws humerus dorsally and posteriorly
Trapezius: A flat superficial muscle extending from the thoracic vertebrae and occipital portion of the skull to the scapula	Occipital bone, cervical and thoracic vertebrae	Spine of scapula	Elevates shoulder

Name and Description	Origin	Insertion	Action
Deltoideus: A thin band of muscle extending from the posterior scapula to the proximal end of the humerus	Aponeurosis of Infraspinatus muscle	Proximal humerus	Adducts humerus
Supraspinatus: A muscle deep to the Deltoideus, posterior to the dorsal aspect of the anterior deep Pectoral; covering the portion of the scapula superior to the scapular spine	Scapular spine and anterior portion of scapula	Proximal humerus	Extends shoulder
Infraspinatus: A muscle deep to the Deltoideus, occupying the infraspinous fossa of the scapula	Lateral, posterior portion of scapula	Proximal, lateral humerus	Abducts and rotates forelimb outward

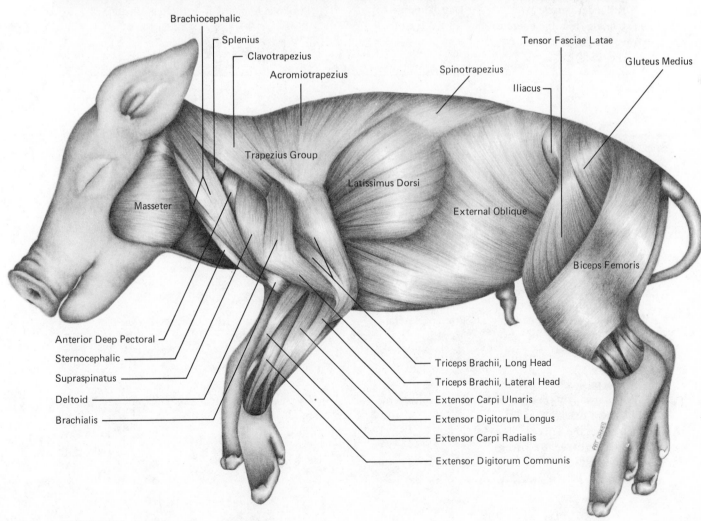

FIGURE 56 *Lateral aspect of fetal pig, superficial muscles*

Name and Description	Origin	Insertion	Action
Rhomboideus: An elongated muscle, consisting of several separate slips; deep to the insertion of the Trapezius. The cephalic portion (Rhomboideus capitis) lies lateral to the cervical and thoracic parts of this muscle and originates at the occipital bone	Occipital bone, cervical and thoracic vertebrae	Dorsal scapula	Draws scapula dorsally
Splenius: A thick muscle deep to the Rhomboideus	Thoracic vertebrae 3, 4, and 5	Superior nuchal line of occipital bone	Elevates head and neck
Serratus Ventralis: Slips of muscle deep to the Latissimus dorsi	Cervical vertebrae and lateral surfaces of ribs	Dorsal, medial scapula	Raises thorax
Brachiocephalic: A slender muscle extending from the humerus to the dorsal aspect of the skull	Mastoid process, Nuchal crest	Proximal humerus	Inclination and extension of neck

MUSCLES OF THE UPPER ARM (See Figure 57)

Name and Description	Origin	Insertion	Action
Triceps Brachii Longus: An elongated muscle covering the posterior surface of the humerus	Posterior border of scapula	Olecranon process of ulna	Extends elbow
Triceps Brachii Lateralis: A band of muscle covering the lateral surface of the humerus	Proximal lateral humerus	Olecranon process of ulna	Extends elbow
Triceps Brachii Medialis: A muscle lying deep to the Triceps brachii lateralis; transect the Triceps lateralis to see the Triceps medialis	Shaft of humerus	Olecranon process of ulna	Extends elbow

Name and Description	Origin	Insertion	Action
Biceps Brachii: A slender, poorly-developed muscle anterior to the Triceps brachii medialis	Ventral portion of scapula anterior to glenoid fossa	Proximal radius and ulna	Flexes elbow
Brachialis: A elongated muscle anterior to the Biceps brachii	Proximal humerus	Proximal radius and ulna	Flexes elbow

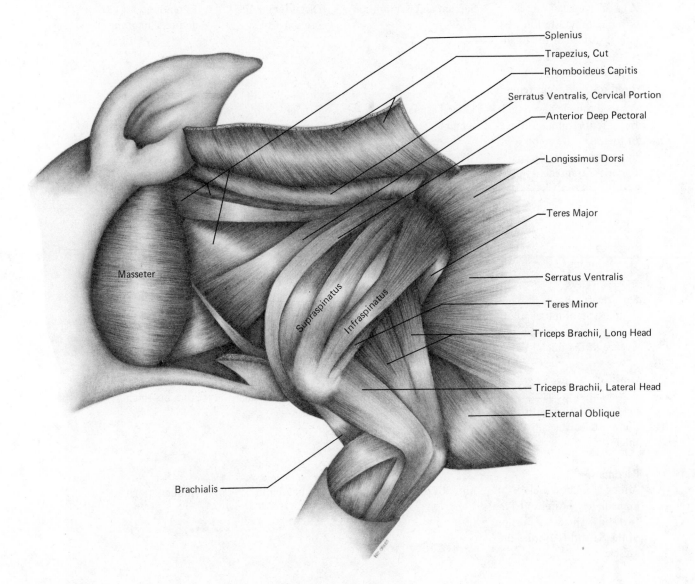

FIGURE 57 *Lateral aspect of fetal pig, left forelimb musculature*

MUSCLES OF THE HIP AND THIGH—DORSAL ASPECT (See Figure 58)

Name and Description	Origin	Insertion	Action
Tensor Fasciae Latae: A broad, triangular muscle extending from the ilium to just superior to the patella	Crest of ilium	Fascia lata	Tightens fascia lata, extends thigh
Gluteus Medius: A thick muscle extending from the ilium to the greater trochanter	Lumbodorsal and gluteal fascius	Greater trochanter femur	Abducts thigh
Biceps Femoris: A large triangular muscle covering the posterior lateral surface of the thigh	Sacrum and ischium	Distal femur and proximal tibia	Extends and abducts hindlimb
Vastus Lateralis (head of Quadriceps femoris): A broad muscle covering the anterior, lateral surface of the thigh. It can be seen more easily if the Biceps femoris is transected	Lateral aspect of femur	Patella	Extends knee

MUSCLES OF THE THIGH—VENTRAL ASPECT (See Figure 55)

Name and Description	Origin	Insertion	Action
Sartorius: A thin strip of muscle on the anterior half of the ventral surface of the thigh	Iliac fossa and tendon of Psoas minor muscle	Proximal, medial tibia	Flexes hip and adducts thigh
Gracilis: A wide, thin muscle covering the posterior half of the ventral surface of the thigh	Ventral surface of pubis	Proximal, medial tibia	Adducts thigh
Pectineus: A long, slender muscle immediately posterior to the Sartorius. The distal portion of the Sartorius overlies this muscle	Anterior pubis	Medial femur	Adducts thigh
Adductor: A long, slender muscle lying deep to the Gracilis and posterior to the Pectineus	Ventral pubis and ischium	Distal medial femur	Adducts thigh

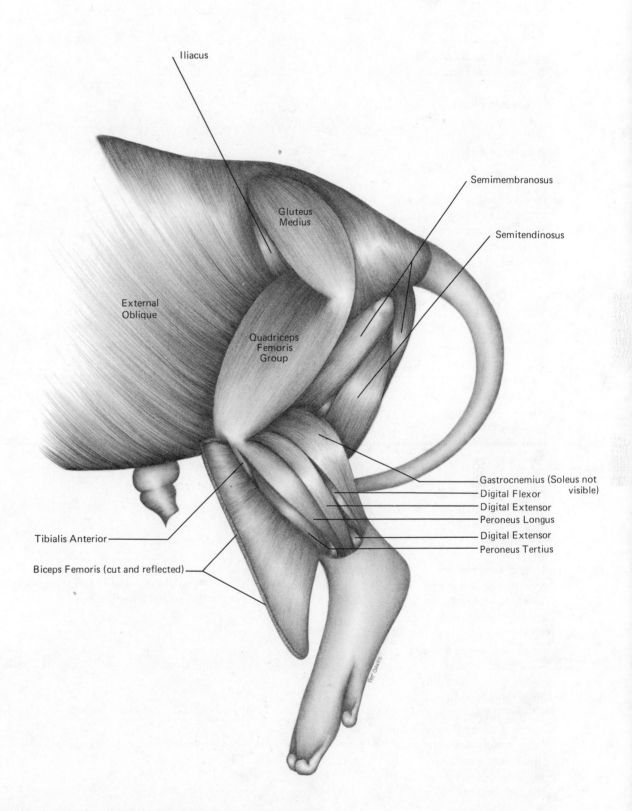

Iliacus

Semimembranosus

Semitendinosus

Gluteus
Medius

External
Oblique

Quadriceps
Femoris
Group

Gastrocnemius (Soleus not
visible)

Digital Flexor

Digital Extensor

Peroneus Longus

Digital Extensor

Peroneus Tertius

Tibialis Anterior

Biceps Femoris (cut and reflected)

FIGURE 58 *Lateral aspect of fetal pig, left hind limb muscles*

Name and Description	Origin	Insertion	Action
Rectus Femoris (head of Quadriceps femoris): A thick muscle lying along the ventral surface of the femur	Anterior ilium	Patella	Extends knee
Semimembranosus: A flat band of muscle deep to the Gracilis; its distal end overlies the Adductor	Ischial tuberosity	Distal femur; proximal tibia	Extends hip and adducts hindlimb
Semitendinosus: This muscle can also be seen on the dorsal aspect of the thigh; it is posterior to the Semimembranosus, and has two heads	First and second caudal vertebrae; ischial tuberosity	Proximal, medial tibia; fascia of leg	Extends hip

MUSCLES OF THE SHANK (See Figure 58)

Name and Description	Origin	Insertion	Action
Gastrocnemius: The calf muscle; has two short, thick heads	Distal end of femur	Calcaneus	Extends ankle
Soleus: A slender, thin muscle lying beneath the Gastrocnemius	Head of fibula	Calcaneus	Extends ankle
Tibialis Anterior: A slender muscle immediately overlying the anterior surface of the tibia	Proximal lateral tibia	Second tarsal and metatarsal	Flexes ankle
Extensor Digitorum Longus: A thin, elongated muscle closely approximated to the Tibialis anterior	Distal lateral femur	Divided tendon to each digit	Extends digits
Peroneus Group (longus, tertius): Thin, elongated muscles extending along the lateral and anterior surfaces of the tibia, respectively	Proximal lateral tibia, distal lateral femur	Tarsals	Flexes ankle

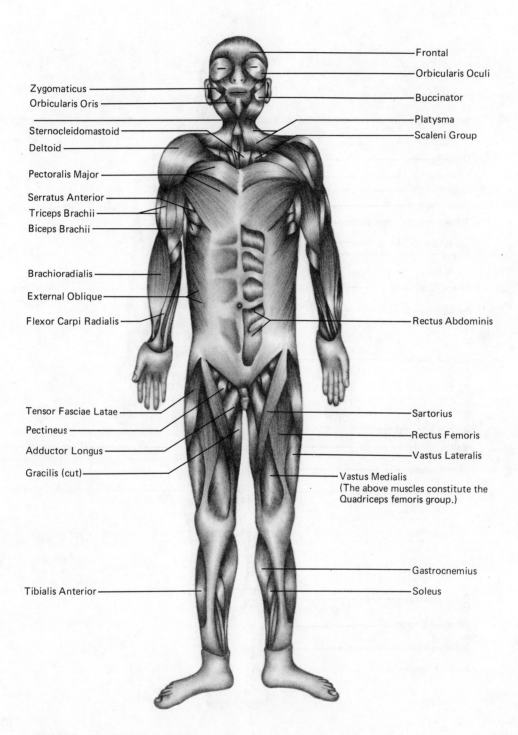

Frontal

Orbicularis Oculi

Zygomaticus

Orbicularis Oris

Buccinator

Platysma

Sternocleidomastoid

Scaleni Group

Deltoid

Pectoralis Major

Serratus Anterior

Triceps Brachii

Biceps Brachii

Brachioradialis

External Oblique

Flexor Carpi Radialis

Rectus Abdominis

Tensor Fasciae Latae

Pectineus

Adductor Longus

Gracilis (cut)

Sartorius

Rectus Femoris

Vastus Lateralis

Vastus Medialis
(The above muscles constitute the
Quadriceps femoris group.)

Gastrocnemius

Soleus

Tibialis Anterior

FIGURE 59 *Human musculature, anterior view*

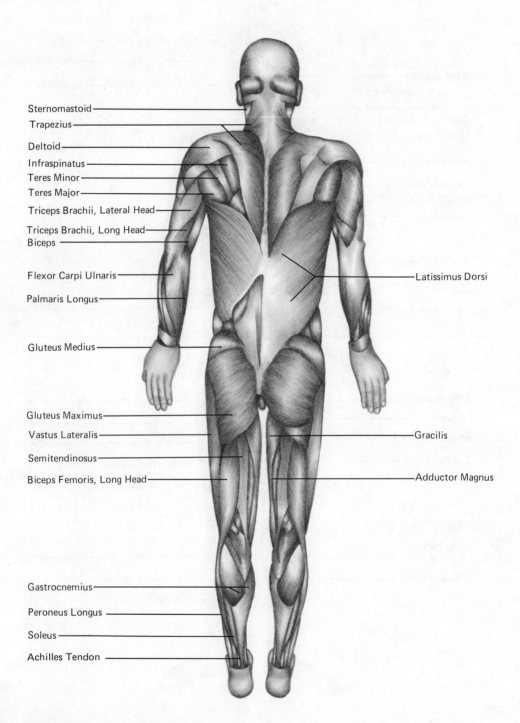

Sternomastoid

Trapezius

Deltoid

Infraspinatus

Teres Minor

Teres Major

Triceps Brachii, Lateral Head

Triceps Brachii, Long Head

Biceps

Flexor Carpi Ulnaris

Palmaris Longus

Gluteus Medius

Gluteus Maximus

Vastus Lateralis

Semitendinosus

Biceps Femoris, Long Head

Gastrocnemius

Peroneus Longus

Soleus

Achilles Tendon

Latissimus Dorsi

Gracilis

Adductor Magnus

FIGURE 60 *Human musculature, posterior view*

EXERCISE 3   *Microscopic Identification of Muscle Types and Myoneural Junctions*

    a. *Skeletal Muscle*

Skeletal muscle is voluntary striated muscle. Obtain a prepared slide of a longitudinal section of skeletal muscle and observe the **muscle fibers** (see Figure 19). Each fiber (muscle cell) is multinucleated. You should be able to distinguish these nuclei quite easily by their peripheral location. Each muscle fiber contains many **myofibrils**, which are made up of **myofilaments**. Careful observation under high power will reveal striations in the myofibrils. **A bands** (*anisotropic bands*) stain darkly and represent myosin and actin filaments. Between the A bands will be light areas known as **I bands** (*isotropic bands*), which represent actin filaments. Dividing the I bands are **Z bands** or **Z lines**. The segment of a fiber between successive Z bands is known as a **sarcomere**, the unit of structure and function of a muscle fiber.

    Observe a cross section of skeletal muscle. An entire muscle is surrounded by connective tissue known as **epimysium**. The muscle is subdivided into **fasciculi**, which are surrounded by **perimysium**. Look for muscle fibers that are contained in a fascicle in your slide. Under high power, you should be able to see the details of a muscle fiber, including nuclei and areas known as **Cohnheim's areas**, which contain myofibrils. In the space below, sketch a longitudinal and a cross section of skeletal muscle. *Label:* nucleus, A band, I band, and Cohnheim's area.

    b. *Smooth Muscle*

Smooth muscle appears nonstriated under the light microscope (see Figures 17 and 18). It is usually considered involuntary and is present primarily in blood vessels and internal organs. Observe a slide of a longitudinal section of smooth muscle. Notice the "dovetailing" effect of the muscle fibers. In cross section, nuclei of smooth muscle fibers are located more centrally within the fiber rather than at the perimeter, as in skeletal muscle. Draw several smooth muscle fibers. *Label:* Nucleus and cytoplasm.

FIGURE 61 *Motor end plate detail as viewed in skeletal muscle*

c. *Cardiac Muscle*

Cardiac muscle is involuntary striated muscle. The muscle fibers bifurcate and connect with adjacent fibers to form a three-dimensional network (see Figure 20). Nuclei of cardiac muscle are situated deep within the fiber as in smooth muscle. Cardiac muscle can be recognized in longitudinal section on your prepared slide by the presence of **intercalated disks**, which join the ends of cardiac muscle fibers together. The pattern of cross striations of myofibrils and the A and I bands and Z lines are the same as in skeletal muscle. Draw a longitudinal section of cardiac muscle. *Label:* Nucleus, A band, I band, Z line, and intercalated disks.

d. *Myoneural Junction*

This slide represents the junction of a nerve ending and skeletal muscle fibers. *Draw and label:* Nerve ending and muscle (Figure 61).

B. The Physiology of Muscle Contraction

PURPOSE	The purpose of Unit VI-B is to enable the student to understand the physiology of muscle contraction.

OBJECTIVES	In order to complete Unit VI-B, the student must be able to do the following: 1. Name the ions that must be present for normal muscle contraction to occur. 2. Understand the importance of ATP in muscle contractions. 3. Calculate the degree of contraction in a Psoas muscle preparation. 4. Induce and interpret various responses to muscle stimulation by electrical and chemical methods.

MATERIALS	microscope slides
	sharp probes
	Psoas muscle preparation (kit*)
	ATP, KC1, MgCl₂ solutions (from kit)
	mechanical or electronic recording
	    apparatus
		living frogs
		scissors
		forceps
		electronic stimulator and
		    electrodes
		frog Ringer's solution

MATERIALS

microscope slides
sharp probes
Psoas muscle preparation (kit*)
ATP, $KC1$, $MgCl_2$ solutions (from kit)
mechanical or electronic recording
 apparatus

living frogs
scissors
forceps
electronic stimulator and
 electrodes
frog Ringer's solution

PROCEDURE

EXERCISE 1   *Chemistry of Muscle Contraction*

In order for a muscle to contract, **ATP (adenosine triphosphate)** must be split and certain ions must be present.

1. Follow your instructor's directions for the preparation of needed solutions or use the solutions in your prepared kit.
2. Place several glycerinated Psoas muscle fibers in a small amount of glycerol on a microscope slide.
3. Measure the lengths of the muscle fibers in millimeters.
4. Flood the fibers with several drops of the magnesium and potassium solutions containing ATP, potassium chloride, and magnesium chloride.
5. After 30 seconds, measure the length of the fibers again. If there is a difference, calculate the percentage contraction.
6. Repeat Steps 2–5, using (1) ATP alone and (2) potassium and magnesium salts alone.

EXERCISE 2   *Induction of Frog Skeletal Muscle Contraction*

A single muscle contraction is known as a **muscle twitch**. A normal muscle twitch has three phases, as shown in the figure. Point 1 is the point of stimulation of the muscle by electrical, mechanical, or chemical means. Distance 1–2 is the **latent period**, during which electrical

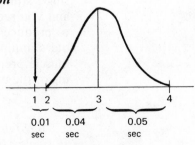

*This kit may be obtained from Carolina Biological Supply Co., Burlington, N.C.

and chemical changes take place within the muscle prior to actual contraction. Distance 2–3 represents the **contraction period**, when the muscle contracts, and Distance 3–4 is the **relaxation period**. Notice that the muscle twitch lasts approximately 0.1 second. The amount of stimulus necessary for a muscle to contract is referred to as the **threshold stimulus** or **liminal stimulus**. A stimulus of less than threshold strength is known as **subthreshold** or **subliminal**.

Your instructor will demonstrate the use of the mechanical or electronic physiological apparatus and the proper procedure for pithing a frog. Pithing involves grasping the frog firmly with one hand holding the head between your thumb and index finger. Incline the head forward about 90 degrees and feel the *occipital condyles* and *foramen magnum* between the condyles with your thumb. Insert a sharp probe into the foramen magnum and anteriorly into the brain. Move the probe from side to side in the brain, then remove the probe. The frog has now been **single pithed** and will feel no pain, the brain having been destroyed.

In order for the frog to lie limp for removal of the nerve–muscle preparation, the spinal cord must also be destroyed. This is done by inserting the tip of the probe into the depression between the occipital condyles and then pointing the probe inferiorly into the spinal cord. Work the probe up and down several times in this region. You should feel the frog relax in your hand. Remove the probe. The frog has now been **double pithed**, which means that both the brain and spinal cord have been inactivated.

Expose the *Gastrocnemius muscle* and **sciatic nerve** in the leg of the frog by cutting the skin around the leg just inferior to the trunk with scissors. Do not cut too deeply. Peel the skin off the leg. From this point, it will be necessary to keep the nerve and muscles moist with frog Ringer's solution. Carefully insert a probe (preferably of glass) under the Gastrocnemius and separate it from the other muscles. With scissors, cut the **Achilles tendon** at the heel. Insert the probe beneath the femur and separate the muscles from around it. Locate the sciatic nerve, which is silver white and stringlike and which should be dorsal to the femur. Sever the sciatic nerve at its proximal end (near the trunk). Remove the nerve–muscle preparation together with the femur. Your instructor will show you how to attach the nerve–muscle preparation to the recording apparatus and will demonstrate the operation of this equipment. *Save your frog* and maintain the intact leg with Ringer's solution for part e. *Muscle Fatigue.*

a. *The Single Muscle Twitch*

In order to cause a muscle to twitch, it is necessary to achieve a threshold stimulus. Setting the voltage knob of the electronic stimulator at its lowest setting, stimulate the muscle itself with the electrodes with a single stimulus. If nothing happens, raise the voltage setting by 5 volts and again apply a single stimulus. Repeat the 5-volt increment until the muscle responds weakly and you see a response on the recording paper. The number of volts needed to produce this weak response is the threshold stimulus. What is the threshold stimulus in this exercise? _____ volts. Remember to keep the nerve–muscle preparation moist with Ringer's solution. *Save and label* the recording of the muscle twitch.

b. *Staircase Phenomenon* (*Multiple Motor Unit Summation* or *Treppe*)

If single stimuli of constant intensity are applied to a muscle, each twitch will be slightly greater than the preceding one. Stimulate the muscle as rapidly as possible with single stimuli for 5 seconds with an intensity of 20 volts above threshold. *LABEL and SAVE* the tracing.

c. *Induction of Muscle Contraction by Chemicals and by Nervous Stimulation*

In this exercise you will be stimulating muscle contraction by using acetic acid and the sciatic nerve.

1. Using the same nerve-muscle preparation, touch the electrode to the sciatic nerve and apply electrical stimuli of increasing intensities, beginning with 0 volts. What is the threshold stimulus?
   _____ volts.

2. Apply several drops of 5% acetic acid to the sciatic nerve. Is there a twitch?
   _____. Repeat if necessary.

3. Apply several drops of 5% acetic acid to the muscle itself, after rinsing the nerve with Ringer's solution.
   Result? _____

4. Rinse the acid off the muscle with Ringer's solution.

LABEL and SAVE the tracings.

d. *Tetanus*

Tetanus is a condition in which muscles are in a state of continual contraction. Setting the voltage at 20 volts above threshold, stimulate the muscle itself in the following manner:

1. A single stimulus at the rate of one per second for 10 seconds. Wait 1 minute.

2. A single stimulus at the rate of two per second for 10 seconds. Wait 1 minute.

3. A single stimulus as rapidly as possible for 10 seconds. This represents *incomplete tetanus.* Wait 1 minute.

4. Stimulate the muscle with multiple stimuli for 10 seconds. This represents *complete tetanus.*

LABEL and SAVE these recordings.

e. *Muscle Fatigue*

Muscle fatigue is caused by the accumulation of waste products including CO_2, lactic acid, and acid phosphates.

1. Determine the threshold stimulus and record several normal muscle twitches.

2. Set the intensity of induction 20 volts above threshold stimulus.

3. With one person watching the time, apply multiple stimuli to the muscle itself, recording the time that elapses between the initial application of stimuli and the beginning of complete tetanus.
   Time: _____.

4. Determine the time between the beginning of complete tetanus and the beginning of fatigue (when the line begins sloping toward the baseline).
   Time: _____.

5. Determine the time it takes for complete fatigue (when the line reaches the baseline).
   Time: _____.

6. Allow the muscle to relax 1 minute, apply Ringer's solution, and then repeat multiple stimulation at 20 volts above threshold. Record times as in Steps 3, 4, and 5. *LABEL and SAVE* the recordings.
   *Times:*_____, _____, _____.
7. Allow the muscle to relax 3 minutes, apply Ringer's solution, and then stimulate and record times as in Steps 3, 4, and 5.
   *Times:*_____, _____, _____.

At this point, the condition of the muscle you are working with may necessitate dissection of the other Gastrocnemius from the frog and using it to replace the first muscle on the recording apparatus. Please remember at all times to keep the muscle moist with Ringer's solution.

f. *Work Performed by Skeletal Muscle*

1. Keep the muscle moist with Ringer's solution. Attach a scale pan to the tendon and determine the threshold stimulus. Then add a 5-gram (5 g) weight to the pan and determine the threshold stimulus.
   _____volts.
2. Set the intensity of induction 100 volts above the threshold stimulus determined above. Do a single muscle twitch. Add additional 5-g weights until the muscle can no longer contract. What weight was needed to achieve this point?
   _____ g.
3. As the weight of the load is increased, there is a proportionate increase in work accomplished to a certain point. After that, an additional load results in less work being performed. The amount of work done by the muscle can be calculated using this formula:

Work = weight of load (in grams) × distance load is lifted (in millimeters).

How much work was done by this Gastrocnemius muscle at its maximum point?
_____ g/mm.

LABEL and SAVE this recording.

Label and save all recordings and attach to Question 5 of the discussion. If this series of exercises was done as a demonstration, then, in the space following Question 5, copy the tracings and hand them in with the other questions. A written interpretation of the various types of muscle contractions accompanying your tracings is required.

Muscular System

DISCUSSION

1. Where is the Psoas muscle located? _____

2. Which ions are necessary for normal muscle contraction? _____
   _____

3. Why is ATP important for muscle contraction? _____
   _____

4. Listed below are several pig muscles. Using reference books, compare them to the human being with respect to presence or absence, location, number of heads, divisions, or size:

   a. Superficial pectoral _____
   _____

   b. Posterior deep pectoral _____
   _____

   c. Anterior deep pectoral _____
   _____

   d. Adductor _____
   _____

   e. Sternocephalic _____
   _____

   f. Sternothyrohyoid _____
   _____

   g. Gluteus medius _____
   _____

5. Attach the following physiological tracings on this sheet. Use additional sheets if needed. Below each recording, write your interpretation of the type of contraction:

   a. single twitch

   b. staircase phenomenon

   c. stimulation of the sciatic nerve by electricity and acetic acid

   d. tetanus

   e. fatigue

   f. work done by skeletal muscle

UNIT VII *Digestive System*

A. *Digestive Anatomy*

PURPOSE

The purpose of Unit VII-A is to enable the student to locate and state the functions of the major digestive organs and accessory glands.

OBJECTIVES

To complete Unit VII-A, the student must be able to do the following:
1. Identify the major divisions and subdivisions of the digestive tract.
2. State the functions of the structures of the alimentary tract.
3. Briefly describe the histology of the digestive tract.

MATERIALS

dissecting instruments
fetal pigs
display material
human skulls

prepared microscope slides of the following: developing tooth, dried tooth, tongue, parotid gland, esophagus, stomach, small intestine, large intestine, appendix, rectum, anus

PROCEDURE

EXERCISE 1   *Digestive System*

The digestive system is comprised of the following divisions and subdivisions:

MOUTH
Teeth
  Incisors
  Canines (cuspids)
  Premolars (bicuspids)
  Molars (tricuspids)
Tongue

PHARYNX
Nasopharynx
Oropharynx
Laryngopharynx

ESOPHAGUS

STOMACH
Cardiac sphincter
Cardiac (diverticulum ventriculi in pig)
Greater and lesser curvatures
Fundus
Pylorus
Pyloric sphincter

SMALL INTESTINE
Duodenum
Jejunum
Ileum

LARGE INTESTINE (COLON)
Ileocecal sphincter
Cecum
Appendix (not present in the fetal pig)
Spiral colon (not present in humans)
Ascending colon
Descending colon
Sigmoid colon (not present in the fetal pig)
Rectum
Anus

ACCESSORY ORGANS AND STRUCTURES OF THE DIGESTIVE TRACT

Pancreas   Common bile duct
Pancreatic duct   Gastrohepatic ligament
Liver   Salivary glands
Gallbladder   Parotid
Cystic duct   Submaxillary
Hepatic duct   Sublingual

Place the fetal pig in the dissecting pan on its dorsal surface. In the study of muscles, you have skinned the pig; therefore, you can now open the abdominal body wall. Do not at any time cut too deeply. The abdominal structures lie immediately under the body wall and can be easily damaged.

Insert the blunt end of your scissors in the midline (linea alba) between the rectus abdominis muscles immediately anterior to the umbilical cord. Gently cut the muscle, proceeding in an anterior direction to the sternum.

If you have not done so, remove the skin posterior to the umbilical cord and observe the penis and scrotal sacs in a male specimen. In general, your specimen must be at least 8 in. long for it to be mature enough for these structures to be differentiated.

Place the fetal pig on its left side in the dissecting pan. If you have not removed the skin on the lateral surface of the head and throat do so now by gently cutting with your scissors. Make your initial incision below the ear and cut in a diagonal plane to the mouth. Be careful not to cut too deeply. The musculature of the fetal pig is poorly developed. The subcutaneous glands and their ducts plus the blood vessels are very close to the skin and are surrounded by connective tissue.

The **parotid gland** may not be well developed but consists of loose glandular tissue beneath the ear in the neck. The **parotid duct** or **Stenson's duct** extends anteriorly and empties into the oral cavity near the maxillary fourth premolar. Beneath the parotid gland and posterior to the angle of the jaw is the **submaxillary gland**. Its duct, **Wharton's duct,** leads into the floor of the oral cavity. Last, identify the **sublingual glands** if possible. They are small and located at the base of the tongue and anterior to the submaxillary glands. The duct also empties into the floor of the oral cavity.

In order to identify the structures within the oral cavity, it will be necessary to cut with bone shears through the jaw at the junction of the mandible and maxilla. Open the cavity further by cutting at the junction with your scalpel. Identify the teeth of the fetal pig. The pig dental formula is:

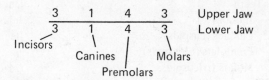

These teeth may not be completely developed.

Identify on a human skull the four types of teeth: the **incisors**, the **canines** (cuspids), the **premolars** (bicuspids), and **molars** (tricuspids).

Identify the **hard** and **soft palates**, the **esophagus**, and **tongue**. The tongue is attached ventrally by the **frenulum linguae** and posteriorly by the hyoid bone. Note the papillae or elevations on the tongue.

The **pharynx** is the throat cavity and is divided into three regions. The area behind the snout is the **nasopharynx**. The region in which the larynx or voice box is situated is referred to as the **laryngopharynx**. The region behind the mouth, known as the **oropharynx,** can be identified by inserting your probe into the **isthmus of the fauces,** the aperture of the throat, and moving the probe anteriorly or posteriorly.

With the blunt end of your scissors continue cutting through the body wall in a posterior direction. *CUT AROUND THE UMBILICAL CORD* and stay *LATERAL* to the midline so that you do not damage the **urinary bladder.**

Continue cutting posteriorly to the pelvis. Make two lateral incisions from the midline cut below the rib cage.

Again, make two lateral incisions from the midline cut at the level of the pubic symphysis. Exercise care not to cut too deeply in this area.

Two flaps, one on either side of the body, away from the midline should have resulted from your dissection.

Observe the smooth glistening lining of the body cavity. This is the **parietal peritoneum** (Figure 62). Also observe the **diaphragm muscle** which separates the **thoracic** from the **abdominal cavity**.

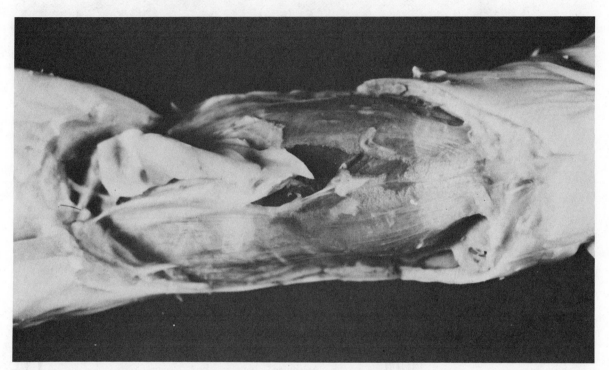

FIGURE 62 *Abdominal muscles reflected to show peritoneum, ventral aspect*

Using Figure 63, locate the **viscera**—that is, all the organs in the abdominal cavity. Note the connective tissue membrane which lies over most of the abdominal organs. This double fold of peritoneum is the **greater omentum**. In order to see the organs beneath it, carefully loosen it and move it to the left side of your specimen.

Following the list of digestive structures in Exercise 1 of this unit, locate those in the abdominal cavity of your specimen.

Trace the **esophagus** where it penetrates the **diaphragm** on the left side and joins with the **stomach** at the **cardiac sphincter**. Examine the external surface of the stomach and identify the **diventriculum ventriculi** region, the **greater** and **lesser curvatures**, the **fundus**, the **pyloric region** and **pyloric sphincter**.

Continue following the **alimentary canal** or **digestive** tract from the **pyloric sphincter** posteriorly to the **small intestine**. Identify the **duodenum, jejunum,** and **ileum** of the small intestine. It will be necessary to gently pull the coiled small intestine free from the dorsal body wall. In order to do this, carefully cut the thin layered **mesentery** (see Figure 63). The mesentery holds the small intestine in its coiled position. By gently cutting the mesentery within the coils, the small intestine can be straightened for easier examination. Be careful not to cut any arteries or veins which are supported by this mesentery.

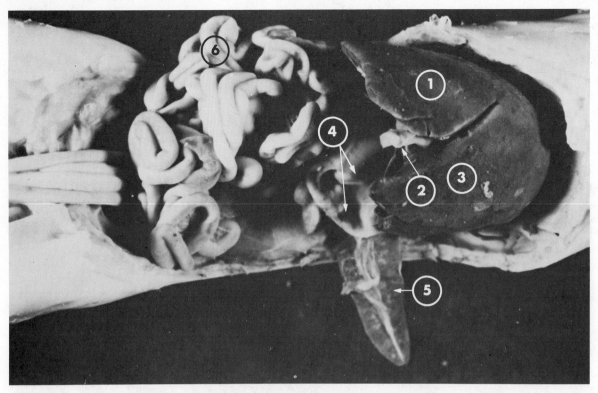

FIGURE 63 *Abdominal muscles and peritoneum reflected to show abdominal viscera, ventral aspect*

1. Liver, Right Medial Lobe   3. Liver, Left Medial Lobe   5. Spleen

2. Umbilical Vein   4. Stomach   6. Small Intestine

FIGURE 64 *Pelvic basin exposed to show ileocecal region, ventral aspect*

1. Duodenum   3. Ileum   5. Ascending Colon   7. Cecum

2. Stomach   4. Descending Colon   6. Ileocecal Valve

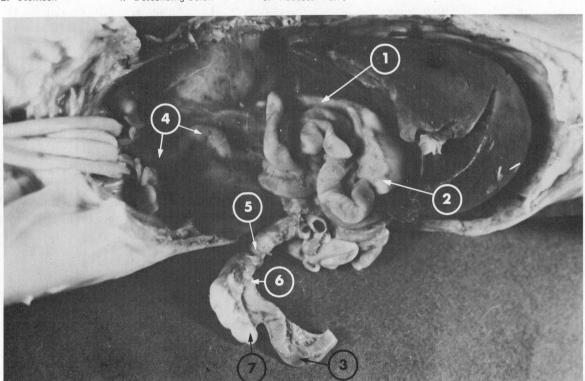

FIGURE 65 *Liver viewed from caudal aspect to show gallbladder*

1. Liver, Left Medial Lobe

2. Hepatic Artery

3. Liver, Left Lateral Lobe

4. Kidney, Left

5. Liver, Right Lateral Lobe

6. Liver, Right Medial Lobe

7. Gallbladder

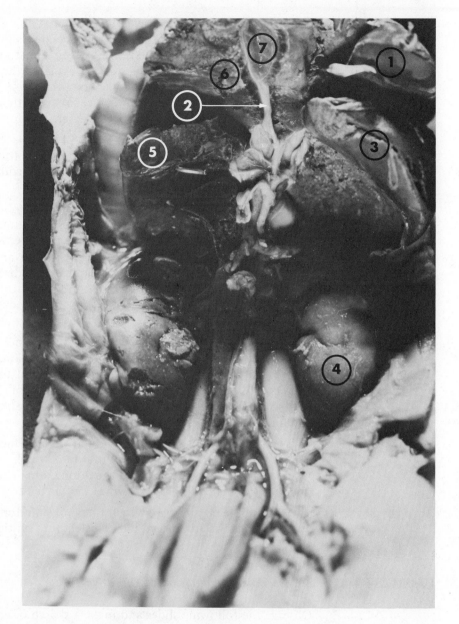

Continue examining the digestive tract by identifying the **ileocecal junction** which joins the **small** and **large intestines.**

The large intestine or colon is divided into subdivisions. The first is the **cecum.** The cecum, a fingerlike pouch, extends outward immediately posterior to the ileocecal junction. The first section of the fetal pig colon is spiral in structure and called the **spiral colon.** In humans this is not true. Also locate the **ascending colon** on the right side and **descending colon** which leads into the **rectum.** In humans the structures anterior to the cecum are the **ascending colon, hepatic flexure, transverse colon, splenic flexure, descending colon,** and **rectum** (see Figure 64).

You have now completed the external examination of the digestive tract and can proceed to an internal examination of the stomach and small intestine. With your scalpel make an incision into the stomach along the greater curvature, from the level of the cardiac sphincter to 1 in. past the pyloric sphincter. Make another incision about 2 in. in length into the jejunum. It may be necessary to flush out these organs for better examination.

Locate the longitudinal wavelike structures lining the stomach. These are the **rugae** which allow the greater surface area and digestion. Examine the **pyloric sphincter,** the constriction between the stomach and small intestine. This band of circular smooth muscle regulates the flow of food from the stomach into the duodenum. In the small intestine examine the **villi.** These minute projections increase the surface area and therefore, the absorption of the end products of digested food. They are not found in either the stomach or colon.

The accessory organs of the digestive system found in the abdominal or **peritoneal cavity** are the **pancreas, liver, gallbladder,** and their associated ducts.

Below the diaphragm note the large purplish, lobular organ, the **liver** (see Figure 65). The fetal pig liver has five lobes, the human liver has four. Identify the **right medial** or **central lobe, right lateral lobe,** and **caudate lobes** of the liver. On the left, identify both the left lateral and left medial lobes. By lifting the right central lobe you can locate and identify the **gallbladder.** The gallbladder is a small saclike organ used for the storage and concentration of bile which is produced by the liver. Observe the **cystic duct** leading posteriorly from the gallbladder.

By probing gently, clear the connective tissue below the left lateral medial lobes. This is the **gastrohepatic ligament.** The **hepatic duct** leading posteriorly from the liver and merging with the cystic duct to form the **common bile duct** should now be obvious. Trace the common bile duct posteriorly and see where it penetrates the duodenum as a papillalike structure termed the **ampulla of Vater.**

The **pancreas** is a very loose glandular organ which lies between the stomach and duodenum. Locate the white duct which exits anteriorly from the pancreas and runs posteriorly to the duodenum. This is the **common pancreatic duct** through which the enzymes are carried to the alimentary canal.

Note the **spleen,** an organ which is responsible for the production of blood cells. It appears as an oblong, reddish structure on the left side of the fetal pig.

EXERCISE 2   *Microscopic Examination of Digestive Tissue*

Observe the following slides and make a sketch of each.

BUCCAL CAVITY

a. *Developing Tooth*
In this slide, the developing deciduous tooth is embedded in the alveolar process of the jaw. Notice the bone of the **alveolus**; the central area of dental **pulp,** which forms the core of the developing tooth; the broad layer of **dentin,** which may stain pink surrounding the pulp; and the **enamel** immediately overlying the dentin. An area known as the *dental sac* surrounds the developing tooth. Also notice the darkly staining *odontoblasts* around the outer margin of the dental pulp. *Sketch and label.*

b. *Dried Tooth*

The broad portion of this section represents the **crown** of the tooth, covered with enamel. The narrower portion represents that part of the tooth beneath the gum line and is referred to as the **root**. Notice that beneath the enamel and extending into the root is a thick layer containing wavy, closely spaced, parallel tubules. This layer is dentin. The junction of the root and crown of the tooth at the gum line is the **neck**. In the region of the root the dentin is covered by a thin layer of **cementum** and a **periodontal membrane**, which is adjacent to the alveolar bone. In the center of your section, you will see a clear area within the dentin. The broader end of this space represents the **pulp cavity**. The more constricted extension of the pulp cavity is the **root canal**. The pulp cavity and root canal contain fibroblasts, odontoblasts, blood vessels, and nerves embedded in connective tissue. *Sketch and label.*

c. *Tongue*

The mucosa of the tongue is stratified squamous epithelium. In some sections, you will be able to see various types of **papillae**, which contain taste buds (see Figures 66–69). Notice the longitudinal, transverse, and oblique planes of skeletal muscle fibers that occupy the interior of the tongue. Embedded in the muscle are lingual glands, blood vessels, and nerves. *Sketch and label.*

d. *Parotid Gland*

Parotid glands, which are one of three types of salivary glands (the others being the **sublingual** and the **submaxillary** or **submandibular** glands), are located inferior and anterior to the ear. The predominant structures on this slide are *serous alveoli,* which secrete a watery saliva. These glands are comprised of pyramid-shaped cells arranged in a circular manner and contain darkly staining nuclei. In this section, there are also blood vessels, adipose tissue, connective tissue, and ducts that drain the glands. *Sketch and label.*

FIGURE 66 *Cleft of papillae showing numerous taste buds*

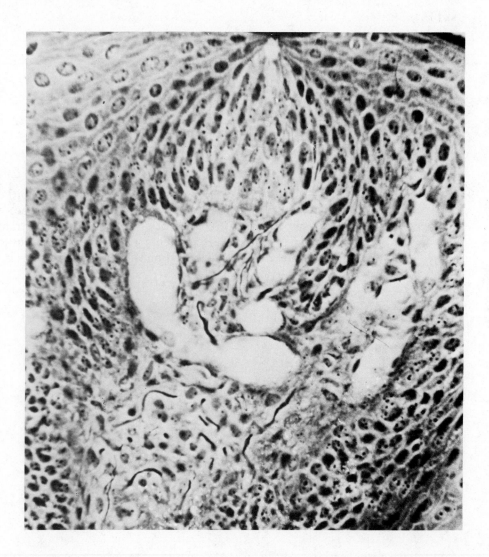

FIGURE 67 *Tongue, showing taste buds*

1. Trench of Vallate Papillum

2. _____

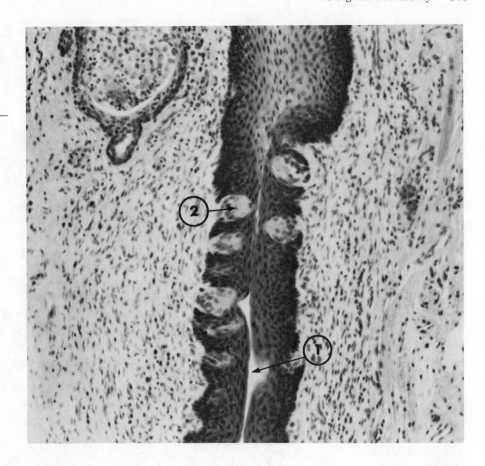

FIGURE 68 *Tongue, filiform papilla*

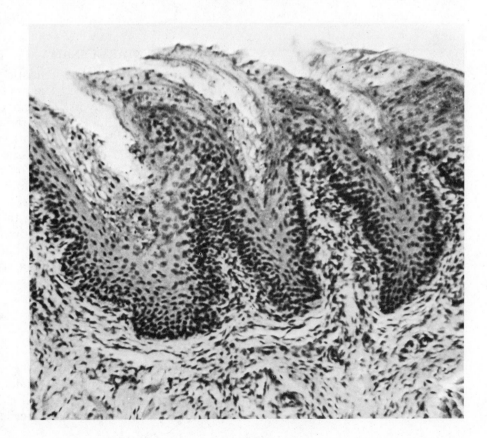

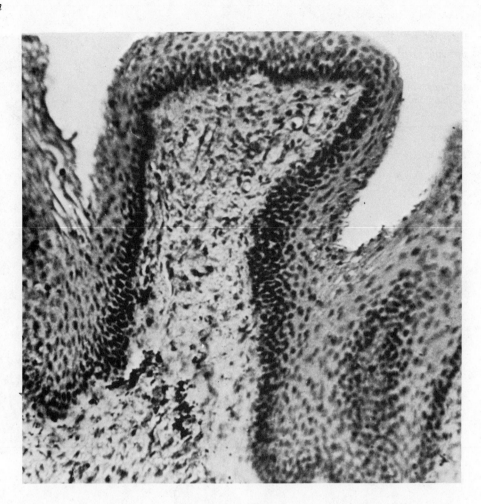

FIGURE 69 *Tongue, fungiform papilla*

ALIMENTARY CANAL

The alimentary canal is essentially a long tube consisting of the esophagus, stomach, small intestine, and large intestine. The wall of the alimentary canal, or digestive tube has four basic layers. The inner lining, or **mucosa,** most often consists of a surface epithelium, an underlying connective tissue layer of **lamina propria** and a thin layer of smooth muscle, **muscularis mucosae.** The second basic layer is a connective tissue layer beneath the mucosa known as **submucosa.** The third layer is the **muscularis externa,** which is usually comprised of two layers of smooth muscle—the inner circular layer and an outer longitudinal layer. The outer layer or coat is **serosa,** which is a reflection of the peritoneum lining the walls of the body cavity. Some parts of the digestive tract have an **adventitia** rather than a serosa. Adventitia is a fibrous outer coat. Modifications of the four basic layers (mucosa, submucosa, muscularis externa, and serosa or adventitia) occur in different regions of the digestive tract (see Figures 70 and 71).

e. *Esophagus*

The mucosa is comprised of stratified squamous epithelium, lamina propria, and muscularis mucosae layers. The submucosa contains *esophageal glands,* which appear similar to serous alveoli of the parotid gland. The remainder of this layer is connective tissue containing adipose cells and blood vessels.

The muscularis externa consists of two layers, the inner one of circular fibers and the outer longitudinal muscle layer. The muscularis externa of the superior third of the esophagus contains striated muscle, the middle third both striated and smooth muscle fibers, and the inferior third smooth muscle only. The outer layer of the esophagus is primarily adventitia, except at the inferior end, where it is serosa. *Sketch and label.*

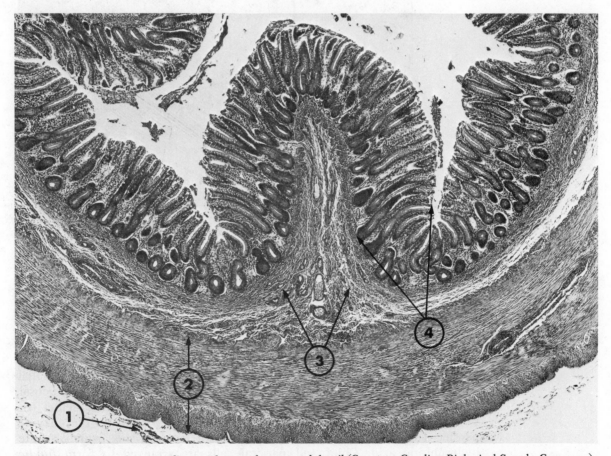

FIGURE 70 *Ileum showing plicae and general structural detail* (Courtesy Carolina Biological Supply Company)

1. Serosa   2. Muscularis Externa   3. Submucosa   4. Mucosa

FIGURE 71 *Stomach, showing detail of mucosa, submucosa, and muscularis layers*

1. _____

2. Muscularis Mucosae

3. _____

4. Mucosa Layer

5. _____

6. Muscularis Externa Layer

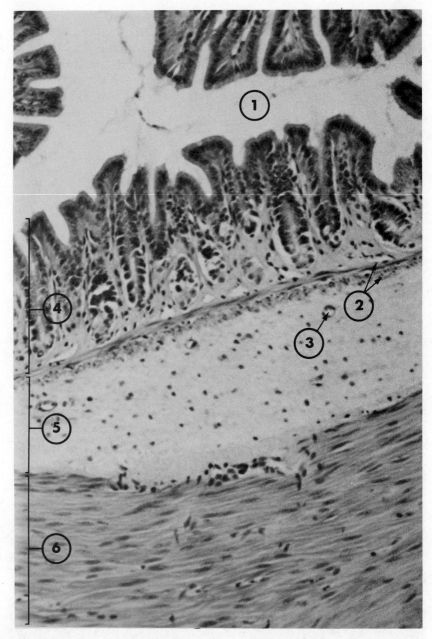

f. Stomach

The stomach has several regions that differ slightly in their histology. The four basic layers are present in all regions and shall be described in general. Observe your section first under low power and then under high power to examine the layers in detail.

The mucosal layer of the stomach consists of a surface of simple columnar epithelium, supported by lamina propria (see Figure 71). If you are observing the fundic or superior portion of the stomach, notice that the epithelial layer dips into the mucosa, forming **gastric pits**. Farther down in the mucosal layer beneath the gastric pits are blue-staining cells, known as **chief** (or *zymogenic*) **cells**, which contain **pepsinogen**, the precursor to the **enzyme pepsin**. There are also cells that stain red, or *parietal cells,* which secrete hydrochloric acid. Chief and parietal cells are arranged to form *fundic glands* (see Figures 72 and 73). Toward the muscularis mucosae you may see some fundic glands in cross section.

The gastric mucosa of the pyloric region of the stomach, which is more proximal to the small intestine, has a different appearance from the fundic region. The gastric pits of the pylorus are deeper, and *pyloric glands* are located between the gastric pits and muscularis mucosae. Cells comprising the pyloric glands stain the same as cells of the gastric pits but are slightly larger in volume. Pyloric gland cells secrete mucus and hydrolytic enzymes. In all areas of the stomach, there are darkly staining lymph nodules at intervals immediately superior to the muscularis mucosae. Lamina propria serves as the connective tissue both in the region of the gastric pits and in the region of the fundic and pyloric glands. The submucosa of the stomach contains blood vessels, adipose, and connective tissue.

The muscularis externa region of the stomach consists of three layers. An inner layer of *oblique fibers* is present in addition to the middle and outer layers of circular and longitudinal fibers, respectively. The serosa is comprised of connective tissue. *Sketch and label* your slide of the stomach.

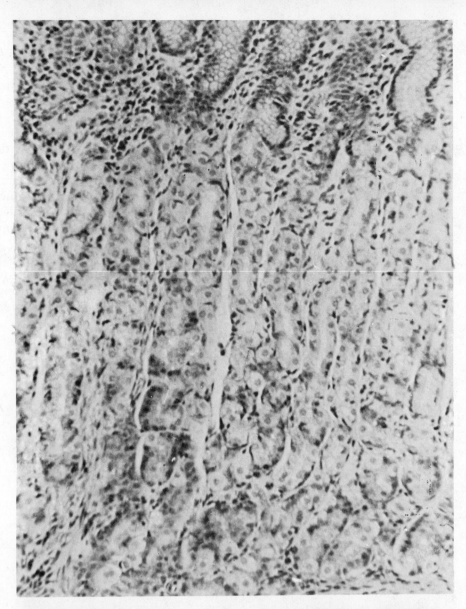

FIGURE 72 *Detail of fundic mucosa of stomach*

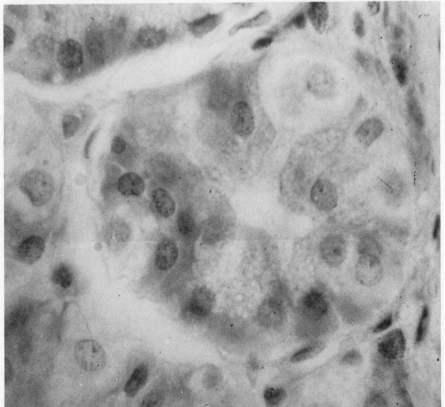

FIGURE 73 *Detail of fundic gland, stomach*

THE SMALL AND LARGE INTESTINES

There are several features common to both the small and large intestines. The mucosa forms deep pits known as *intestinal crypts of Lieberkühn* or **intestinal glands**. The intestinal mucosa contains four cell types, each having a particular function. The frequency with which these cell types appear depends upon the particular region of the intestine. Simple **columnar** epithelial cells line the inner surface of the intestinal mucosa. *Goblet cells* secrete mucus and are thought to be formed through the transformation of simple columnar cells. *Argentaffin cells* (see Figure 74) are usually found between other cells lining intestinal glands. *Paneth cells* are located at the base of intestinal glands. Argentaffin cells have a granular cytoplasm and are thought to secrete *serotonin,* a powerful stimulant of smooth muscle contraction. Paneth cells are pyramidal in shape, with a basophilic cytoplasm that stains darkly and granules that stain with acid dyes, usually red or orange, depending upon the particular stain used. Other common features of the intestinal tract mucosa are lamina propria, which is found underlying the mucosal epithelium and between intestinal glands; lymph nodules, which are more prominent in certain parts of the intestine than others; and a thin muscularis mucosae, which has an inner layer of circular and an outer layer of longitudinal smooth muscle fibers.

FIGURE 74 *Detail of goblet cells from a jejunum villus*

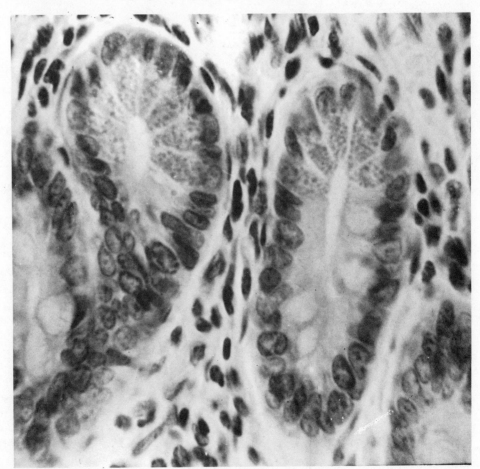

The submucosa of the intestinal tract consists of loose areolar tissue, adipose, blood vessels, and nerves. There may be other structures present in the submucosa, depending on the particular intestinal region. The muscularis externa consists of an inner circular smooth muscle layer and an outer longitudinal layer. The outer layer of the intestinal tract is serosa.

g. *Small Intestine*

The small intestine is divided into three regions: the **duodenum**, which is approximately 10–12 inches long; the **jejunum**, which is about 8 feet long; and the **ileum**, which is about 12 feet long. The entire small intestine is characterized by the presence of **plicae circulares** (or *valves of Kerckring*), which are folds in the mucosa and submucosa, and by **villi** (see Figure 70), which are microscopic fingerlike projections of the mucosa overlying the plicae. Villi further increase the absorptive surface area of the small intestine (see Figure 75). Crypts of Lieberkühn can be seen inferior to the villi and superior to the muscularis mucosae.

Observe a section of duodenum. The primary histological characteristic of this region is the presence of *duodenal glands* or **Brunners's glands**, which are found primarily in the submucosa (see Figure 76) and, to some extent, in the mucosa. The secretions of these glands are of a mucoid nature. Also notice the four basic layers in this section: the mucosa, with mucosal goblet cells, argentaffin cells, and simple columnar cells, as well as the crypts of Lieberkühn, and muscularis mucosae; submucosa; muscularis externa; and serosa (see Figure 77). *Sketch and label.*

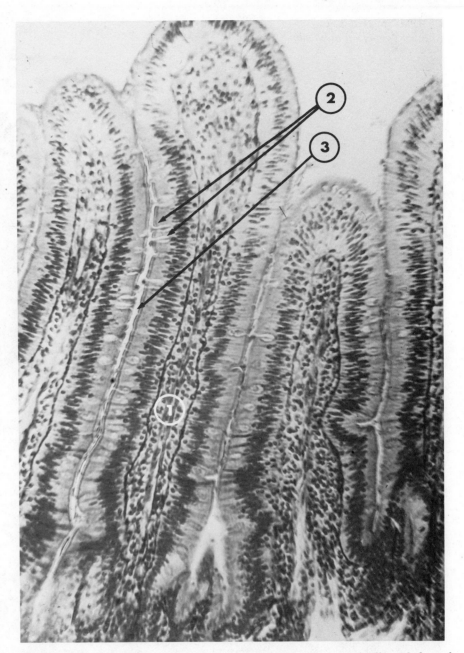

FIGURE 75 *Longitudinal section showing general structure of villi with lacteal detail visible*

1. Lacteal
2. Goblet Cells
3. Striated Border

FIGURE 76 *Detail of Brunner's glands of duodenum*

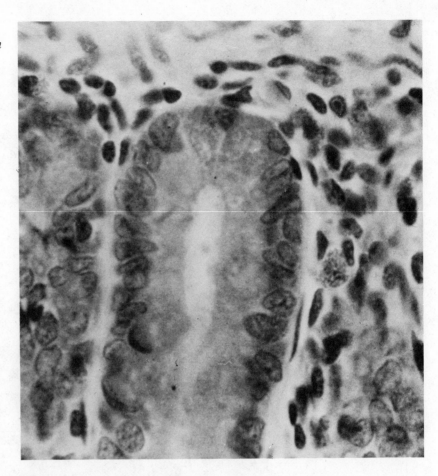

FIGURE 77 *Cross section of jejunum showing muscular wall, villi, and glands* (Courtesy Carolina Biological Supply Company)

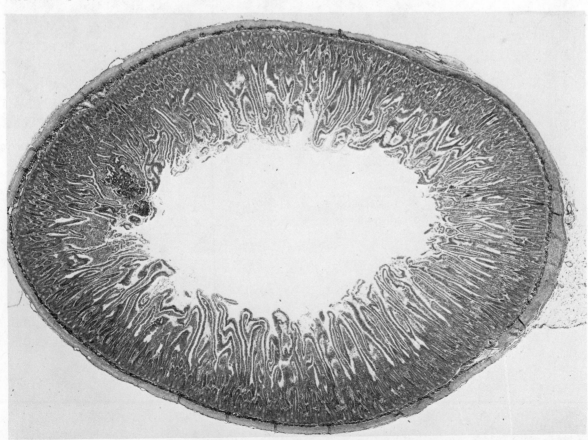

The sections of the jejunum and ileum are similar. Notice the presence of darkly staining lymphatic nodules in the submucosa. In the ileum, aggregations of these nodules are known as **Peyer's patches**. Also notice that the villi become more elongated as one progresses from the duodenum to the jejunum to the ileum. Goblet, Paneth, and simple columnar cells should be visible on these sections (see Figures 78 and 79). *Sketch and label.*

h. *Large Intestine*

The large intestine lacks villi (see Figure 80). Intestinal glands or crypts in the mucosa are very prominent. Notice the numerous goblet cells (see Figure 81) in the lining of the intestinal glands in this section. *Solitary lymph nodes* can be seen in the lamina propria of the mucosa, immediately adjacent to the muscularis mucosae. Also observe the submucosa, muscularis externa, and serosa in this section. *Sketch and label.*

i. *Appendix*

Notice the poorly developed intestinal glands and numerous lymph nodules in this section. *Sketch and label.*

j. *Rectum*

The rectum is characterized by the presence of large longitudinal folds of mucosa and submucosa known as *rectal columns.* These columns are seen in transverse section on your slide. Notice the four basic layers here, including the crypts of Lieberkühn in the mucosa. *Sketch and label.*

k. *Anus*

The mucosa of the anus is stratified squamous epithelium. The lamina propria contains a plexus of large veins, which when dilated are known as *hemorrhoids. Sketch and label.*

FIGURE 78 *Microvilli, detail of small intestine* (Courtesy Carolina Biological Supply Company)

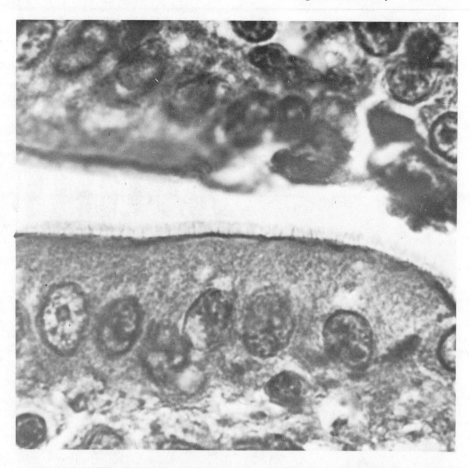

FIGURE 79 *Detail of ganglion cells of Auerbach's plexus* (Courtesy Carolina Biological Supply Company)

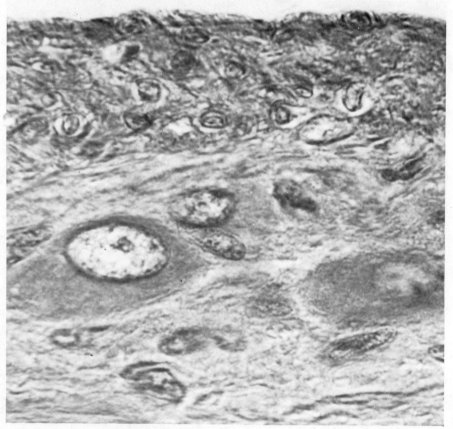

FIGURE 80 *General structure of colon*

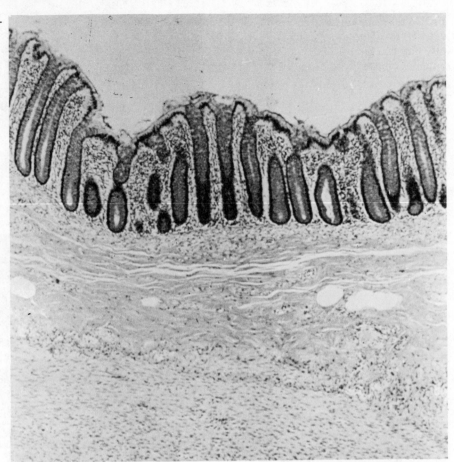

FIGURE 81 *Colon, mucosa layer*

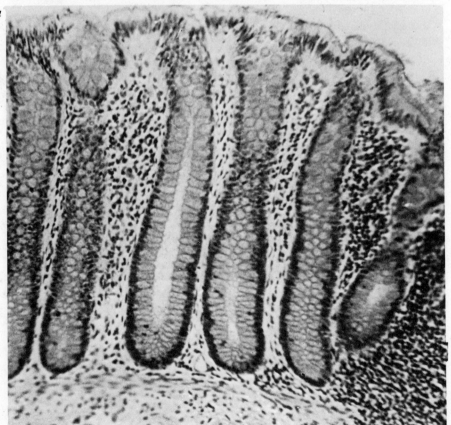

B. *Digestive Chemistry*

PURPOSE The purpose of Unit VII-B is to enable the student to understand the basic chemical principals of digestion and introduce to the student the nature and scope of enzymatic activity which are indicative of metabolic reactions within all cells.

OBJECTIVES In order to complete these exercises, the student must be able to do the following:

1. Summarize the digestion of proteins, carbohydrates, and fats.
2. Perform chemical tests for the identification of various amino acids, proteins, carbohydrates, and fats.
3. Summarize enzyme activity in terms of pH, temperature, and concentration ratios.

SUMMARY OF PROTEIN, CARBOHYDRATE, AND FAT DIGESTION

Digestion of Protein

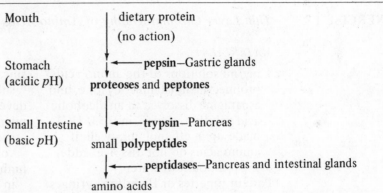

Mouth   dietary protein (no action)

Stomach (acidic *p*H)   ← **pepsin**—Gastric glands
proteoses and **peptones**

Small Intestine (basic *p*H)   ← **trypsin**—Pancreas
small **polypeptides**

← **peptidases**—Pancreas and intestinal glands
amino acids

Digestion of Carbohydrate

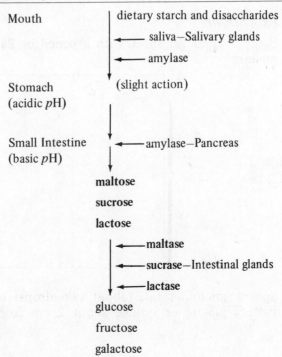

Mouth   dietary starch and disaccharides
← saliva—Salivary glands
← amylase

Stomach (acidic *p*H)   (slight action)

Small Intestine (basic *p*H)   ← amylase—Pancreas

maltose

sucrose

lactose

← **maltase**
← **sucrase**—Intestinal glands
← **lactase**

glucose

fructose

galactose

Digestion of Fat

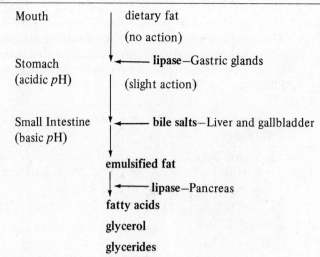

Mouth   dietary fat
               (no action)

Stomach   lipase—Gastric glands
(acidic *p*H)
               (slight action)

Small Intestine   bile salts—Liver and gallbladder
(basic *p*H)

               emulsified fat

                   lipase—Pancreas
               fatty acids
               glycerol
               glycerides

EXERCISE 1   *Thin Layer Chromatography of Amino Acids*

MATERIALS

1 mg/ml solutions of the amino acids proline, phenylalanine, cystine, and asparagine dissolved in an alcoholic acid solution, that is, in 0.5*M* HC1 made up in ethanol; mix well; if amino acids do not dissolve, add a slight amount of water

Pasteur pipettes or Hamilton syringes

concentrated ammonium hydroxide solution

60°C oven

silica gel thin layer chromatography sheets

developing tank

migrating solvent: butyl alcohol: acetic acid: water   80:20:20 volume/volume

ninhydrin solution: 0.3% ninhydrin in butyl alcohol containing 3% glacial acetic acid; or commercially prepared ninhydrin spray

PROCEDURE

Score a silica gel sheet with a pencil or Pasteur pipette in the following manner:

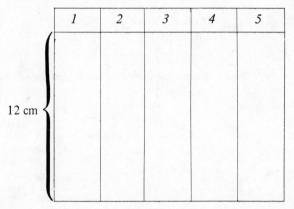

Spot 1 microliter (μl) (about two drops) of amino acid solution with a Pasteur pipette or syringe about 2 cm from the bottom of the sheet in

columns 1–4, applying a different amino acid in each column. Be sure to keep a record of which amino acid you spotted in each column. In the fifth column, spot an amino acid taken from a vial labeled "unknown," which actually contains one of the four amino acids you are using.

Keep the spot in each column as small as possible, blowing gently to facilitate drying. Allow the spots to dry 15 minutes. To neutralize the acid, hold each of the five spots over the open mouth of a bottle of concentrated ammonium hydroxide for a few seconds. Without touching the surface of the gel, place the sheet into a developing tank containing the migrating solvent. Cover. Develop the chromatogram for approximately 3 hours or until the solvent has traveled a distance of 12 cm (10 cm above the spots). Place a maximum of two sheets in each tank.

When the migration is complete, carefully remove the sheet from the tank, mark the solvent front (distance the solvent traveled up the sheet), and set aside until dry. Spray the developed sheet with 0.3% ninhydrin or dip it in a ninhydrin solution. Place the chromatograms on a tray and heat in a 60°C oven for several minutes until the separated zones appear clearly visible. Do not place the chromatogram directly on the oven shelf.

Most amino acids will appear purple, but some will be brown or yellow. Measure the distance traveled by each amino acid and by your "unknown." Calculate an R_f (ratio of fronts) for each column:

$$R_f = \frac{\text{distance traveled by spot (in centimeters)}}{\text{distance traveled by solvent front (in centimeters)}}$$

The R_f value is always a value between 0 (spot did not move) and 1 (spot moved with the solvent front). Each amino acid has a different R_f value.

Record the appearance of spots on your plate in this space:

1	2	3	4	5

R_f value of each spot   Name of amino acid

1. _____   1. _____

2. _____   2. _____

3. _____   3. _____

4. _____   4. _____

5. _____   5. _____

Identify your unknown in column 5 by comparing its R_f value with the R_fs of the known amino acids in a chart provided by your instructor.

EXERCISE 2   *Tests and Properties of Proteins*

Proteins are compounds comprised of amino acids joined together with peptide bonds:

$$
\underset{\text{glycine}}{\underset{\displaystyle\text{NH}_2}{\overset{\displaystyle\text{H}}{\text{H—C—COOH}}}} + \underset{\text{alanine}}{\underset{\displaystyle\text{NH}_2}{\overset{\displaystyle\text{H}}{\text{H}_3\text{C—C—COOH}}}} \xrightleftharpoons{-\text{H}_2\text{O}} \underset{\text{dipeptide}}{\underset{\displaystyle\text{NH}_2\ \ \text{H CH}_3}{\overset{\displaystyle\text{H O H}}{\text{H—C—C—N—C—COOH}}}} + \text{H}_2\text{O}.
$$

All amino acids and proteins contain carbon, hydrogen, oxygen, and nitrogen. A few proteins also contain other elements.

MATERIALS

1% albumin solution	0.1% freshly pre-	ring stands
or egg white	pared ninhydrin	5% trypsin solution
10% NaOH	undiluted egg white	Bunsen burners
1% CuSO$_4$	distilled water	test tube holders
Millon's reagent	test tubes	test tube racks
0.1% alcohol solution	beakers	medicine droppers
of glycine or alanine		

PROCEDURE

a. *Biuret Reaction*

The **Biuret reaction** is specific for compounds containing two or more peptide bonds. To 3.0 cc of 1% albumin solution or egg white, add 3.0 cc of 10% NaOH and one drop of 1% CuSO$_4$. Mix. Add additional drops of CuSO$_4$ until a violet color is obtained, indicating a positive reaction.

b. *Millon Reaction*

The **Millon reaction** is specific for the amino acid tyrosine, which contains a benzene ring to which is attached a hydroxyl (–OH) group. Add a few drops of Millon's reagent to 5 cc of 1% albumin solution. Boil carefully. A brick-red color indicates the presence of tyrosine, a common amino acid in most proteins.

c. *Ninhydrin Reaction*

The **Ninhydrin reaction** is a test for alpha amino acids. To 5 cc of 0.1% alcohol solution of glycine or alanine, add 0.5 cc of 0.1% ninhydrin (freshly prepared). Heat. Note color formation. Repeat, using 3 cc of 1% albumin solution instead of an amino acid. Heat to boiling the albumin to which ninhydrin was added and cool. *Color?* _____

d. *Heat Coagulation*

Place 5 cc of undiluted egg white in a test tube and place in a water bath. Boil. *Describe your results. Is this reaction reversible upon cooling and agitation?*

e. *Digestion of Protein*

Place a small piece of cooked egg white from Exercise 2(d) above into each of two test tubes. To Tube 1, add 10 ml distilled water (buffered to pH 8). To Tube 2, add 5% trypsin or pancreatin solution. Allow to stand in a warm place for several hours. Observe. *Describe your results:*

Record the results of Exercises 2(a)–2(e) in the table in Question 9 of the Discussion following this unit.

EXERCISE 3   *Tests and Properties of Carbohydrates*

All **carbohydrates** contain the elements carbon, hydrogen, and oxygen, the last two of which are present in the same ratio as in water (2:1). Carbohydrates may be classified as mono-, di-, or polysaccharides. The **mono-** and **disaccharides** resemble each other considerably, but polysaccharides bear little resemblance to the other two classes.

MATERIALS

4% glucose solution	Barfoed's reagent	beakers
5% alcoholic solution of	4% fructose solution	test tubes
alpha naphthol	Seliwanow's reagent	medicine droppers
concentrated H_2SO_4	$1M$ HCl	ring stands
Benedict's solution	$1M$ NaOH	test tube holders
1% glucose solution	thin starch paste	test tube racks
1% sucrose solution	Lugol's iodine	

PROCEDURE

a. *The Molisch Reaction*

The **Molisch reaction** is a general test for carbohydrates. In a test tube, place 5 cc of 4% glucose solution. Add two drops of 5% alcoholic solution of alpha naphthol. Mix. Place 5 cc concentrated sulfuric acid (*CAREFUL*) in a second tube. With the tube containing glucose in your left hand, pour the sulfuric acid (H_2SO_4) slowly and gently down the side of the tube containing glucose. The object is to get a layer of H_2SO_4 under the glucose solution. If a purple or pink color is formed at the boundary, the substance is a carbohydrate. *Result?*

b. *Benedict's Test*

Benedict's test is a test for reducing sugars, which include glucose, fructose, galactose, mannose, maltose, and lactose. To 5 cc of Benedict's solution, add four to five drops of 1% glucose solution. Boil in a water bath for 2 minutes. Cool slowly. A red, green, or yellow precipitate (ppt) indicates a positive reaction. Repeat, using sucrose. *Results?*

c. *Barfoed's Test*

Barfoed's test is a test for monosaccharides. To 5 cc of Barfoed's reagent, add five to six drops of 4% glucose solution. Boil in a water bath for 1 minute. Set aside for 15–20 minutes and watch for a red ppt, which indicates a positive reaction. *Results?*

d. *Seliwanow's Test*

Run this test in duplicate, using one tube with glucose and another tube with fructose. Add five drops of 4% glucose solution to one tube containing 5 cc Seliwanow's reagent and five drops of 4% fructose to another. Boil for 30 seconds. What does **Seliwanow's test** enable you to do?

e. *Inversion of Sucrose*

Add 5 cc of 1% sucrose solution to each of two test tubes. To the first tube, add 5 cc Benedict's solution and bring to a boil. Remove from heat immediately. To the second tube, add two drops of dilute (1M) HCl and boil in a water bath for 1 minute. Neutralize with an equal amount and strength of NaOH. (Why?) Test this solution with 5 cc Benedict's solution. How do you explain the different results for these two procedures?

f. *Test for Starch and Starch Hydrolysis*

Place 5 ml of thin starch paste into a test tube. Add one drop of Lugol's iodine solution. Is there a color change?_____ Put 10 ml of the same starch solution into a test tube and add four or five drops of saliva. Set in a 37°–40°C water bath for ½ hour or more. Is there a color change when iodine is added? Explain.

Record the results of Exercises 3(a)–3(f) in the table in Question 9 of the Discussion following this unit.

EXERCISE 4   *Tests and Properties of Fats*

Fats are the esters (organic salts) formed by the union of a fatty acid and an alcohol. The three common fats in our foods and body are oleic, palmitic, and stearic, which are formed, respectively, from oleic, palmitic, and stearic acids united chemically with the alcohol glycerol (glycerin). One molecule of glycerol can unite with three fatty acids:

$$H_3C-(CH_2)_{16}-\overset{\overset{O}{\|}}{C}-OH$$

$$H_3C-(CH_2)_{16}-\overset{\overset{O}{\|}}{C}-OH \; + \; \begin{matrix} HO-CH_2 \\ HO-CH \\ HO-CH_2 \end{matrix} \; \underset{+3H_2O}{\overset{-3H_2O}{\rightleftharpoons}}$$

$$H_3C-(CH_2)_{16}-\overset{\overset{O}{\|}}{C}-OH$$

3 molecules of stearic acid   1 molecule of glycerol

$$\begin{matrix} H_3C-(CH_2)_{16}-\overset{\overset{O}{\|}}{C}-O-CH_2 \\ H_3C-(CH_2)_{16}-\overset{\overset{O}{\|}}{C}-O-CH \\ H_3C-(CH_2)_{16}-\overset{\overset{O}{\|}}{C}-O-CH_2 \end{matrix} \; +3H_2O.$$

tristearin; (a triglyceride)

MATERIALS

fresh cottonseed oil	sweet cream
distilled water	0.5% Na_2CO_3
ethyl alcohol	test tubes
ether	test tube racks
5% trypsin	Bunsen burners
beakers	thermometers
benzene	ring stands
carbon tetrachloride	blue litmus solution

PROCEDURE

a. *Solubility of Fat*

Set up five test tubes as follows. First, make sure that there are no Bunsen flames burning in the area in which you are working.

Tube No.	Solvent (5 ml)
1	water
2	ethyl alcohol
3	ether
4	benzene
5	carbon tetrachloride

Add five drops of fresh cottonseed oil to each tube and shake well. Observe tubes immediately and again after 15 minutes. *Result?*

b. *Digestion of Emulsified Fat*

1. Add 10 ml sweet cream to a test tube.
2. Add 8 to 10 drops of 1% blue litmus solution (as indicator) to the cream to impart a blue color. Mix.
3. Pour 5 ml of this mixture into another test tube.
4. Add 2 ml of 5% trypsin to one tube and 2 ml 0.5% Na_2CO_3 to the second tube.
5. Place both tubes into a beaker of 40°C water. Observe after 1 hour. Did you observe a color change? Explain.

Record the results of Exercises 4(a) and 4(b) in the table in Question 9 of the Discussion following this unit.

EXERCISE 5   *Enzymes: Chemical Activity in Living Systems*

The purpose of this experiment is to introduce the student to the nature and scope of biochemical activities which are indicative of metabolic reactions within all cells. The liver contains enzyme systems, one of which is catalase, the enzyme that will be studied in this experiment.

MATERIALS

The following materials are sufficient for each group of four (4) students:

(1) 5-cc disposable syringe
(1) Ring stand
(1) Buret clamp
(1) 50-ml buret
(1) 18 in. of aquarium tubing
(8) 50-ml Erlenmeyer flask
(1) 1(one)-hole rubber stopper
(1) 3-in. glass tubing (¼ in. O.D.)
(1) Water trough
       Stock Liver Extract Solution—20%, 50%, 70%, (500 ml each/30 students)
(1) Wax marking pencil
       Blender (for entire class)
       3% H_2O_2 (520 cc/30 students)
       Ringer's solution (1480 cc/30 students)
       Buffer solutions (pH 2, 5, 7, 8, and 12) of 100 ml quantity.
       Liver (fresh baby beef) 2 lbs/30 students
(2) Wood splints
       Ice
(1) Finger bowl
(1) °C thermometer
       Paper towels

A. APPARATUS SETUP PROCEDURE

1. Several students should begin to assemble the apparatus as shown in Figure 82, while the other partners prepare the liver extract. Fill the water trough three-fourths full with tap water. Fill the 50-ml buret to the top with water. Check to see that the stop cock is closed. Place your thumb over the top of the opening of the buret; invert it and submerge in the water trough. Clamp the buret securely and place a paper towel between the clamp and buret in order to avoid breakage.
2. Cut glass tubing to approximately 3 in. and insert in a twisting motion into the rubber stopper until 2 in. are below the stopper.
3. Attach the aquarium tubing securely to the outside portion of the glass tubing and set this aside. The apparatus should now appear as shown in Figure 82.

B. LIVER EXTRACT PREPARATION

While your partners prepare the apparatus, you may now help to prepare the liver extract. *SEVERAL STUDENTS MAY PREPARE EXTRACT FOR THE ENTIRE CLASS.*

1. Take one-half pound of fresh beef liver, dice it, and place in a blender. Add 500 ml of Ringer's solution into the blender and blend until the consistency of a very thin malt. When completed, pour into a flask or beaker and label *Stock Liver Extract*. This stock solution will be the basis for determining the effects of temperature and substrate concentrations.
2. Solutions of liver extract must now be prepared with the following buffered pH solutions: 2, 5, 7, 8, and 12 (others may be substituted).

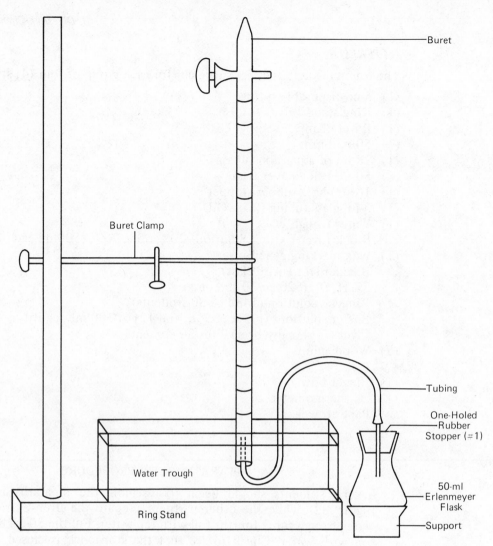

FIGURE 82 *Enzyme apparatus*

Use one-quarter pound of liver diced and chopped. Place in blender and add 330 ml of buffer of pH 2. Blend until the consistency of a thin malt. Pour into a flask and label *Liver Extract pH 2.* Rinse out the blender container with distilled water and repeat the previously mentioned procedure for all the other pH solutions.

3. You have now prepared all of the *Liver Extract Stock Solutions* which will be utilized during the experiment. You will need to make dilutions of the *Stock Liver Extract Solutions* which you prepared in Step 1. Use the following dilutions and label:

Concentration Ratios	Add
20%	8 ml of Ringer's solution and 2 ml of *Stock Liver Extract*
50%	5 ml of Ringer's solution and 5 ml of *Stock Liver Extract*
70%	3 ml of Ringer's solution and 7 ml of *Stock Liver Extract*

The enzyme solutions have now been prepared. These solutions will be utilized later in the experiment in order to demonstrate the effects of enzyme substrate concentration.

C. EXPERIMENTAL PROCEDURES

You have now completed the task of setting up the apparatus and producing the *Stock Liver Extract* at the various concentrations and pH values. The only other variable that must be considered will be that of the effects of temperature on the *Stock Liver Extract Solution.* This variable will be adjusted at the time of use.

1. Place exactly 10 ml of *Stock Liver Extract Solution* into the 50-ml Erlenmeyer flask. Place the one-holed stopper with the glass fitting securely on the top of the flask and insert the opposite end of the aquarium tubing into the open end of the buret which is submerged in the water. You may have to shift the buret in order to position the tubing correctly. Water will go into the aquarium tubing. Check to make sure that all connections of tubing are tight before proceeding to Step 2.

2. Obtain a disposable syringe and draw 1 cc of hydrogen peroxide (H_2O_2) into the barrel of the syringe.

3. Insert the syringe needle into the aquarium tubing rapidly at the junction of the tubing and the glass fitting. Rapidly push the plunger down the syringe and quickly remove from the tubing.

4. Now shake the Erlenmeyer flask with a slow and constant rotary motion.

5. In order to determine the rate of this reaction, watch the amount of water in milliliters which is displaced into the water trough. Observe the bubbles of gas over a period of 3 minutes. Begin this 3-minute timing the instant you introduce the hydrogen peroxide (H_2O_2).

6. Slowly turn the buret valve releasing the accumulated gas into an inverted test tube. (Do not permit the water to enter the buret valve!) Quickly insert a glowing wood splint into the test tube and check the gas.
   a. *Hydrogen*—it will whistle loudly as it ignites and the test tube will turn warm to the touch.
   b. *Oxygen*—it will ignite the glowing splint into flame.
   c. *Carbon dioxide*—it will smother the glowing splint.

What kind of gas have you collected? _____

7. *Repeat steps 1 through 5,* since you are probably unfamiliar with this procedure and have introduced errors into the experiment. It is important that you perform the previously mentioned steps accurately throughout the remainder of the experiment. You have now seen the rate of reaction for 100% substrate enzyme concentration. This will be the basis for comparing the rest of the parameters that we will investigate; namely; temperature, pH, and varying enzyme concentrations.

8. *Record* the ml of H_2O displaced in 3 minutes under the *100% Enzyme Stock Solution (Control)* headings in Questions 10–12 of the discussion.

D. TEMPERATURE INVESTIGATION

1. Use a clean, dry 50-ml Erlenmeyer flask and at the end of each experiment rinse out the flask with distilled water and dry it.

2. The 20°C room temperature is to be used first in investigating the amount of oxygen that will be produced by the enzyme stock solution. Record your results in the table in Question 10 of Discussion at the end of this unit.

3. Place 10 ml of *Stock Enzyme Solution* into a clean 50 ml Erlenmeyer flask and place this flask into a finger bowl which has cold water and ice cubes in it. Obtain a clean Centigrade thermometer. Place a one-holed

      stopper into top of the flask with the aquarium tubing attached to the stopper.

4. Check to make sure that the opposite end of the aquarium tubing is in the open end of the buret and that this is completely submerged in the water. Double check to make sure that all connections of tubing are tight.

5. Obtain the disposable syringe and draw 1 cc of hydrogen peroxide (H_2O_2) into the barrel of syringe.

6. Insert the syringe needle into the aquarium tubing rapidly at the junction of the tubing and the glass fitting. Rapidly push the plunger down and quickly remove from the tubing.

7. *Immediately shake the Erlenmeyer flask with a slow and constant rotary motion.*

8. In order to determine the rate of this reaction, watch the amount of water in milliliters which is displaced into the water trough. Observe the bubbles of gas over a period of 3 minutes. Begin the 3-minute timing in the instant you introduce hydrogen peroxide (H_2O_2).

9. Repeat the previous procedures for 37°C and 42°C temperatures. Place water into the finger bowl. Adjust the hot and cold faucets to produce a temperature that is 42°C. Now place the Erlenmeyer flask into the finger bowl containing 42°C water. Wait 1 or 2 minutes and then repeat Steps 3 through 8. Repeat using 45°C water in the finger bowl. Record your results in the table in Question 10 of Discussion.

10. Again record the amount of gas displacement in milliliters that is present within the buret. *NOTE:* IF THE WATER LEVEL IS LOW IN THE BURET, THEN REFILL THE BURET SO THAT THERE IS NO AIR WITHIN IT.

E. pH INVESTIGATION

Earlier, you prepared *Liver Extract Solutions* with various buffers and you labeled these *Liver Extract pH−*. The pH solutions of 2, 5, 7, 8, and 12 (substitution may be made by your instructor) will be the ones that we will investigate in terms of the effect pH has on enzyme substrates.

1. Use five clean, dry Erlenmeyer flasks and label them 2, 5, 7, 8, and 12, respectively.

2. Take 10 ml of the *Liver Extract–pH 2* and place this volume in a labeled flask. Investigate each of the pH values.

3. Place the stopper and glass tubing arrangement into your Erlenmeyer flask labeled pH 2 and check to make sure all tube fittings are tight and that there is sufficient water within the buret.

4. Obtain the disposable syringe and draw 1 cc of hydrogen peroxide (H_2O_2) into the barrel of the syringe.

5. Insert the syringe needle into the aquarium tubing at the junction of the tubing and the glass fitting and rapidly push the plunger down and then quickly remove the syringe.

6. *Immediately shake the Erlenmeyer flask while you observe for 3 minutes.*

7. Again, observe the rate of this reaction over a period of 3 minutes. Begin this 3-minute timing the instant you shake the flask and upon introduction of the hydrogen peroxide with the syringe.

8. Record in the table the amount of water displacement.

9. Repeat the previous procedure for pH 5, 7, 8, and 12.

10. Record all results in the table in Question 11 of the Discussion at the end of this unit.

F. ENZYME CONCENTRATION INVESTIGATION

1. Earlier, you prepared from the *Stock Liver Extract Solution,* concentrations which were labeled 20%, 50%, and 70%. You will use these to investigate the amount of enzyme necessary to produce the enzyme/ substrate reaction.

2. Place 10 ml of the 20% *Liver Extract Solution* into a clean 50 ml Erlenmeyer flask. Then place 10 ml of the 50% and 70% solutions into the other Erlenmeyer flasks. Label each flask.

3. Again take the disposable syringe and insert 1 cc of hydrogen peroxide (H_2O_2) into the system.

4. Gently shake the flask for the 3-minute observation period.

5. Again observe the rate of this reaction by the amount of water in milliliters which is being displaced by the gas from this reaction.

6. Record in the table the amount of water in milliliters that was displaced by the gas from this reaction.

7. Repeat the above procedures for the 50% and 70% concentration ratios and record in the table in Question 12 of the Discussion at the end of this unit.

Digestive System

DISCUSSION

1. Compare the dental formulae of the pig and human being.

2. Why is it important to have increased surface area in the small intestine?

3. How do the lobes of the liver in the human being differ from the pig?

4. Compare the anatomy of the colon in the pig and human being with respect to the presence or absence of:

   a. appendix

   b. spiral colon

   c. hepatic flexure

   d. sigmoid colon

5. In which portion of the digestive tract does the most digestion take place? _____

   _____

6. What is meant by a "reducing" sugar? _____

   _____

7. Which sugar is used as table sugar? _____

8. Which digestive glands secrete the following?

   a. amylase _____

   b. pepsin _____

   c. trypsin _____

   d. peptidase _____

   e. lactase _____

   f. lipase _____

131

9. Summarize your results of the protein, fat, and carbohydrate exercises in this table:

Test	Result	Conclusion
1. Biuret reaction		
2. Millon reaction		
3. Ninhydrin reaction		
4. Heat coagulation of protein		
5. Digestion of protein		
6. Molisch reaction		
7. Benedict's test		
8. Barfoed's test		
9. Seliwanow's test		
10. Inversion of sucrose		
11. Starch test		
12. Starch hydrolysis test		
13. Solubility of fat		
14. Digestion of emulsified fat		

10. Record the results of temperature investigation in this table.

100% Enzyme Stock Solution (Control)	ml H_2O Displaced in 3 Minutes	ml H_2O Displaced in 1 Minute
20°C		
37°C		
42°C		

11. Record the results of your investigation of the effect of pH on enzyme substrates in this table.

pH Values	ml H_2O Displaced in 3 Minutes	ml H_2O Displaced in 1 Minute
2		
5		
7		
8		
12		
100% Enzyme Stock Solution (Control)		

12. Record the results of your investigation of enzyme concentrations necessary to produce a reaction in this table.

	ml H_2O Displaced in 3 Minutes	ml H_2O Displaced in 1 Minute
20%		
50%		
70%		
100% Enzyme Stock Solution (Control)		

13. Graph the results you obtained, using separate ordinates for temperature, pH, and concentration results. For the X-axis, use the parameter that you investigated; and for Y-axis, use the rate of enzyme reaction which you determined.

14. After graphing, indicate the optimum activity in terms of pH, temperature, and enzyme concentration. Indicate the minimum enzyme activity in terms of pH, temperature, and enzyme concentration. _____

_____

15. Briefly explain why there is an optimum point for enzyme activity within the investigated parameters. Is there any proportionality (inverse or direct) related to increasing or decreasing of the parameter investigated? _____

_____

16. During a viral infection, afflicted individuals may have temperatures up to 104°F over an extended period of time. What happens to the enzyme systems within human cells? How are they affected by this increase in body temperature? _____

_____

UNIT VIII *Circulatory System*

A. *Anatomy of Blood, Blood Vessels, and Heart*

PURPOSE

The purpose of Unit VIII-A is to enable the student to identify the blood vessels, anatomical features of the heart, characteristics of blood, and patterns of systemic and pulmonary circulation.

OBJECTIVES

To complete Unit VIII-A, the student must be able to do the following:

1. Observe the differences in the microscopic structure of an artery, vein, capillary, and lymph vessel.
2. Identify microscopically the components of normal and abnormal human blood and frog blood.
3. Become familiar with the location and function of the major blood vessels in the fetal pig with reference to the human system.
4. Identify the structures of the pig and sheep heart.
5. State the functions of arteries, veins, capillaries, lymph vessels, blood, and the heart.

MATERIALS

compound microscope
prepared microscope slides of artery, vein, capillary, lymph vessels,
   normal human blood smear (Wright's stain), sickle-cell anemia,
   frog blood, and leukemia

EXERCISE 1

Microscopic Examination of Artery, Vein, Capillary, and Lymph Vessels

The circulatory system is designed to transport blood to and from the capillaries. An exchange of vital respiratory gases and other metabolic substances occurs between the capillaries and the cells.

Arteries carry blood from the ventricles of the heart to the capillaries in progressively smaller vessels called *arterioles* and *metarterioles*. Arteries vary in cross section from large vessels, such as the aorta, to smaller tubes known as arterioles. The arterial wall is composed of three distinct layers: (1) the **tunica intima**, or innermost layer, consists of a thin layer of endothelium resting on an *internal elastic membrane;* (2) the **tunica media**, or middle layer, is composed of a layer of thick smooth muscle fibers and elastic connective tissue; and (3) the **tunica adventitia**, or outermost layer, is made up of loose collagenous fibers and adipose tissue. Larger blood vessels contain minute vessels termed **vasa vasorum** in the tunica adventitia, which nourish the walls of larger arteries and veins. Because of the elastic fibers and thick wall of the artery, these vessels do not readily collapse during periods of low blood pressure and are capable of expanding to accommodate high blood pressures.

Capillaries are thin-walled single-layered vessels. Through this endothelial layer, the exchange of oxygen, carbon dioxide, and other metabolic substances occurs. Squamous epithelium together with loose connective tissue composes the endothelium of capillaries.

Blood flows into the veins after passing through a capillary network. Veins, like arteries, are composed of three layers; however, the layers are thinner. The *tunica media* is composed of several layers of smooth muscle; the *tunica adventitia* is a well-developed layer of collagenous fibers, which indicates that connective tissue predominates in veins. Veins, such as those in the extremities, contain **semilunar valves** in their walls assisting the return of blood to the heart against the force of gravity. The valves consist of semilunar flaps, which open to permit the flow of blood to the heart and close to prevent the back flow of blood. Lymphatic vessels contain valves serving the same purpose.

PROCEDURE

1. Examine a prepared slide of a cross section through an artery under low power. Identify the three layers of this vessel.
2. In order to see the detail of the arterial wall, observe it under high power. Observe the **endothelium** of the *tunica intima* layer. Note the wavy *internal elastic membrane* that separates the tunica intima from the

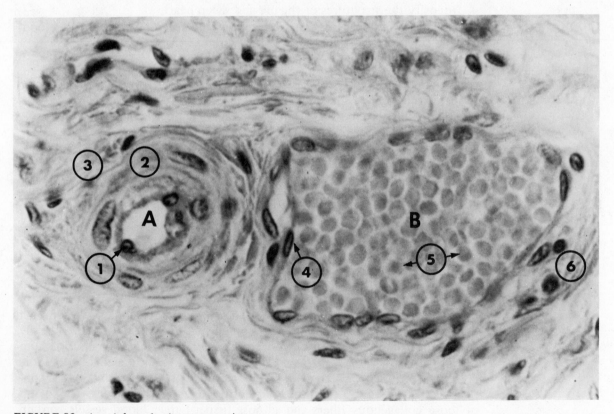

FIGURE 83 *Arteriole and vein, cross section*

A. Arteriole Lumen

1. Endothelial Cell
2. Tunica Media, Muscle
3. Tunica Adventitia

B. Venule Lumen with Blood Cells

4. Endothelial Cell
5. Blood Cells
6. Tunica Adventitia

tunica media. Locate the tunica media layer, which contains thick smooth muscle and elastic connective tissue. Identify the *external elastic membrane* that separates the tunica media from the tunica adventitia. Try to locate the vasa vasorum. *Draw and label* in detail a small section through the arterial wall in Box A (Labeled Drawing of Artery, C.S.) following this exercise.

3. Examine a prepared slide of a cross section through a vein under low and high power. Identify the *tunica intima, tunica media,* and *tunica adventitia* layers of the vein. Note the differences between the arterial and venous walls. *Draw and label* in detail a small section through the venous wall in Box B following this exercise.

4. Examine under high power a prepared slide of a capillary network. Note the single layer of endothelial cells of the capillary wall that are separated by a connective tissue membrane (see Figure 83). *Draw* in Box C provided following this exercise.

5. Examine under low power a prepared slide of a longitudinal section of a lymphatic vessel containing a valve (see Figure 84). Lymphatic vessels contain numerous valves that are more closely spaced than those found in veins. The valves have their free margin orientated in the center of the vessel and have a swollen beaded appearance. *Draw* in Box D following this exercise.

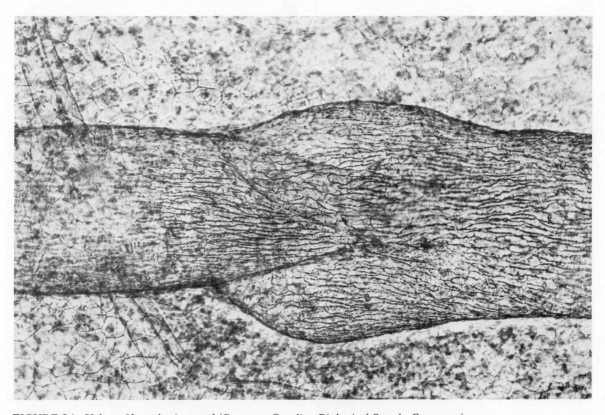

FIGURE 84 *Valve of lymphatic vessel* (Courtesy Carolina Biological Supply Company)

A. *Labeled Drawing of Artery (cross section)*

B. *Labeled Drawing of Vein (cross section)*

C. *Drawing of Capillary Network*

D. *Drawing of Lymphatic Valve*

EXERCISE 2   *Microscopic Identification of Prepared Blood Slides*

Blood is a specialized form of connective tissue and consists of formed and fluid elements. Blood is transported through vessels to receive oxygen from the lungs and collect metabolic products from the alimentary tract. Other substances are also contained in the blood in order that they may be transported in the blood stream to all parts of the body.

Fresh whole blood, when centrifuged, separates into two distinct portions—fluid and formed—known as *elements*. The fluid portion is a clear, straw-colored fluid known as **plasma**. The plasma fraction of the blood accounts for more than 50% of blood volume. The formed elements consist of red blood cells, **erythrocytes**; white blood cells, **leukocytes**; and **platelets** or *thrombocytes*.

The *erythrocytes* are highly specialized cells for the transportation of oxygen. Each cell is biconcave, and mature cells lack a nucleus (enucleate). Erythrocytes are elastic and are capable of considerable distortion as they pass through capillary walls. In human males, there are 5,000,000 red blood cells per cubic millimeter, and in females, 4,500,000–5,000,000. If a sample of blood is spread thinly on a microscope slide and stained properly with Wright's blood stain, the erythrocytes stain red. Erythrocytes have a tendency to adhere to one another along their concave surfaces, a phenomenon known as **Rouleaux formation**.

White blood cells, or *leukocytes,* are cells containing various shaped nuclei. Leukocytes average 5000–9000 per cubic millimeter in normal blood. The count in children is higher and identifiable variations from the normal number occur in pathological conditions. Different types of leukocytes may be identified in a differential white blood cell count. Leukocytes are classified as *agranular* and *granular. Agranular leukocytes* have a homogeneous cytoplasm and large round nuclei. **Lymphocytes** and *monocytes* are agranular. *Granulocytes* can be distinguished by the presence of cytoplasmic granules and polymorphic nuclei. The granular leukocytes are of three types: **acidophils** (or **eosinophils**), **basophils**, and **neutrophils**.

Agranular leukocytes include:

1. *Lymphocytes:* These leukocytes have a relatively large, round, dark-staining nucleus surrounded by a narrow rim of clear blue cytoplasm. Lymphocytes may be classified as small, medium, or large. They constitute 25%–30% of the leukocytes of normal blood (see Figure 85).
2. *Monocytes:* These are larger than lymphocytes and have an oval or kidney-bean-shaped nucleus surrounded by a pale grayish-blue cytoplasm when stained with Wright's stain. These cells constitute 3%–8% of the leukocytes of normal blood (see Figure 86).

Granular leukocytes include:

1. *Neutrophils:* These are the most numerous leukocytes and constitute 65%–75% of the total white blood cell population. The neutrophil is a **polymorphonuclear** leukocyte—that is, the nucleus shows a variety of forms but usually consists of from three to five irregularly oval-shaped lobes connected by thin strands of nuclear material. This is sometimes referred to as the "drumstick" nucleus. The cytoplasm is filled with fine, pale lilac granules (see Figure 87).
2. *Acidophils* (*Eosinophils*): These cells are somewhat larger than neutrophils. The nucleus is usually irregularly shaped and partially constricted into two lobes. The cytoplasm is filled with coarse, round, uniformly sized red-orange granules. They constitute about 2%–4% of the normal white blood cell count (see Figure 88).

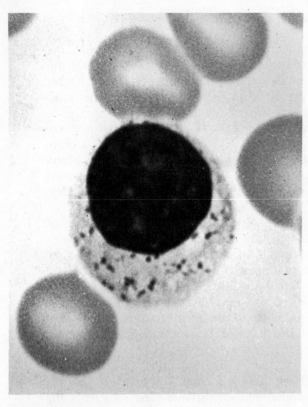

FIGURE 85   *Lymphocyte*

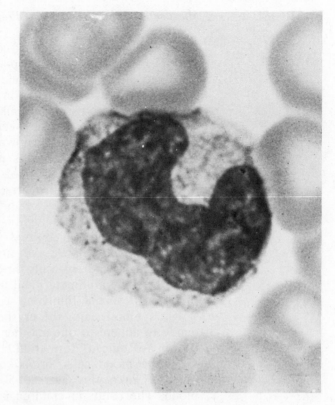

FIGURE 86   *Monocyte*

3. *Basophils:* These cells are difficult to find in human blood since they constitute only 0.5%–1.0% of the total number of white blood cells. The nucleus is irregularly shaped and may appear as two lobes. The cytoplasm is filled with irregular blue-purple granules, which usually obscure most of the bilobed nucleus (see Figure 89).

The function of leukocytes is to combat infection. The white blood cells are one of the body's major defense mechanisms against bacteria, viruses, and other foreign substances that have entered the blood stream. Neutrophils and monocytes are termed **phagocytes.** They have the unique ability to phagocytose or engulf foreign substances. As a result, most foreign invading substances are digested and inactivated by this process. At the site of invasion, death of some surrounding tissue takes place. This **necrotic** (dead) mixture gives rise to **pus.** Lymphocytes are all-purpose cells that may be transformed into other white blood cell types. Lymphocytes may in effect become **antibodies.** Eosinophils (acidophils) seem to have the role of functioning in clot dissolution. Basophils secrete the anticoagulant *heparin.* Heparin is a normal constituent of blood and is necessary in order to prevent *intravascular clotting.*

Blood platelets, sometimes referred to as *thrombocytes,* are small disk-shaped structures that circulate freely in blood. The normal platelet count in adults varies between 200,000 and 400,000 per cubic millimeter of blood. Their essential role is **hemostasis.** This is the process by which platelets adhere to injured regions of blood vessels by accumulating to produce a blood clot known as a *thrombus.* Platelets are a factor in forming extrinsic *thromboplastin,* which is important in the clotting mechanism. Tissue thromboplastin (clotting factor III) is derived from extravascular sources.

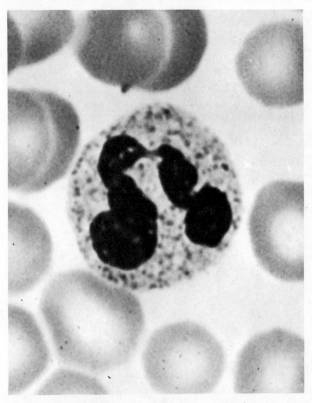

FIGURE 87 *Neutrophil* *red granules pale blue*

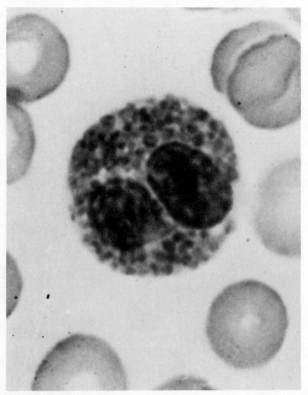

FIGURE 88 *Eosinophil* *~~Blue dk purple granules~~ Red granules*

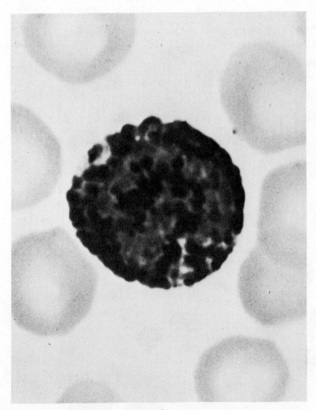

FIGURE 89 *Basophil* *Blue*

FIGURE 90 *Human peripheral blood smear*

1. Platelet
2. Neutrophil
3. Monocyte
4. Erythrocyte

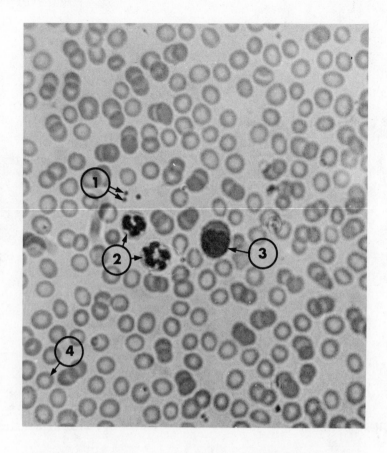

FIGURE 91 *Frog blood*

1. Nucleus of Red Blood Cell
2. Eosinophil

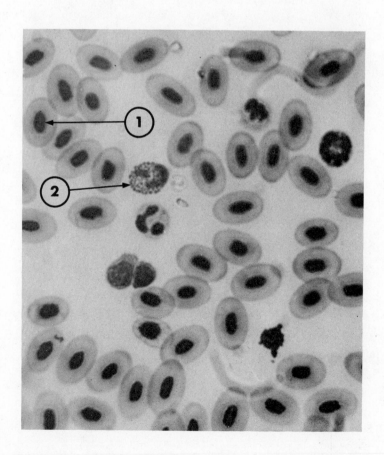

PROCEDURE

A. NORMAL HUMAN BLOOD

Examine, first under the high and dry objective and then under oil immersion, a prepared human blood smear stained with Wright's blood stain. Identify the following cell types: *erythrocytes* (which are enucleate), *monocytes, neutrophils, lymphocytes, basophils, acidophils,* and *platelets.* Basophils and acidophils may be difficult to identify in this preparation (see Figure 90).

B. FROG BLOOD

Examine a prepared frog blood smear for *acidophils* and *basophils,* which appear similar to human cells. Notice that frog erythrocytes are large nucleated cells (see Figure 91).

C. SICKLE-CELL ANEMIA

Notice the sickle-shaped erythrocytes on this slide. These cells have a reduced capacity for carrying oxygen. This is an inherited condition in which the hemoglobin of the abnormally shaped cells contains the amino acid valine substituted for the amino acid glutamic acid, which is present in normal hemoglobin.

D. LEUKEMIA

Leukemia is a malignant condition characterized by an abnormal increase in white blood cells and the presence of immature blood cells in the peripheral circulation. If your slide is of *myelogenous* leukemia, there will be an increase in granular leukocytes. If you have a *lymphogenous* leukemia slide, you will see an increase in lymphocytes in addition to immature cells.

EXERCISE 3   *Dissection of the Fetal Pig Circulatory System*

Your specimen has been properly prepared for the dissection of the arterial and venous blood vessels by the injection of colored latex. This latex simulates the natural texture of the blood vessel. In order to facilitate the identification of blood vessels, the arteries have been injected red and the veins blue. Arteries are injected via the umbilical artery in the umbilical cord. Veins are injected via the jugular vein in the neck.

In order for you to identify blood vessels, all arteries and veins must be teased free from the adjacent supporting tissue for these structures to be clearly visible. One must take great care to avoid severing these vessels while exposing them.

Retract the ribs laterally and using dissecting pins, pin the superior portions of the ribs to the shoulder. Using your blunt metal probe, identify the lungs and trachea within the thoracic cavity before beginning dissection of the circulatory system.

MATERIALS

double-injected and preserved
fetal pig specimens

demonstration specimens
dissecting instruments

PROCEDURE

DISSECTION OF ARTERIES WITHIN THE THORACIC CAVITY
(Figures 92, **foldout preceding Unit I**, and 93)

1. If you have not already done so, remove the **thymus gland**—a tan to brown irregular granulated organ that functions in immunity—from the region anterior to the heart and trachea. The heart is covered with a protective membrane, the pericardium, which is not attached to the heart but to the ventral superior blood vessels. Determine where the pericardium is attached, then slit it open and carefully remove it. The ventral surface of the heart should be exposed.

2. The **aorta** is the largest artery emerging from the heart, which forms the **aortic arch** prior to descending along the dorsal body wall of the thoracic and abdominal cavities. Connective tissue should be carefully removed in order to expose this vessel.

3. The **pulmonary artery** extends from the right ventricle of the heart and carries oxygen-poor blood (deoxygenated) to each lung. These vessels may or may not be injected with blue latex. Follow this blood vessel and note that it branches into the right and left pulmonary arteries that enter each lung. Prior to birth the blood is shunted from the pulmonary circulation into the general systemic patterns. The umbilical vein provides the necessary oxygen and nutrients to the developing fetus prior to birth.

4. Locate the **coronary arteries,** which lie on the ventral surface of the heart. Note the groove—the **coronary sulcus**—that separates the right atrium from the right ventricle. A second groove may be found that separates the right ventricle from the left ventricle. This longitudinal is termed the *anterior longitudinal sulcus.*

Divisions of the Brachiocephalic Artery

5. Locate the **brachiocephalic artery,** the first and largest branch emerging anteriorly from the aortic arch. In the fetal pig and man, this blood vessel first divides into a **right subclavian artery,** a **right common carotid artery,** and a **left common carotid artery.** The second branch off the aortic arch is the left subclavian artery. In the human being, there are three branches off the aortic arch. The first branch is the brachiocephalic (innominate) artery, which divides into the right subclavian and right common carotid arteries. The second branch is the left common carotid artery. The third vessel branching off the aorta is the left subclavian artery.

6. Locate the left and right subclavian arteries on your specimen and observe two small arteries that parallel the common carotid arteries. These are the **thyrocervical arteries** which supply the thyroid gland and pectoral muscles of the chest.

7. At the superior border of the larynx, the left common carotid artery divides into **external** and **internal carotid arteries.** The internal carotid artery is the smaller of the two and travels deep into the occipital region of the skull. The external carotid is much larger and branches many times to supply blood to the head.

8. Return to the **right subclavian artery** at its origin from the braciocephalic. Locate the **vertebral artery,** which emerges from the dorsal portion of the right subclavian artery under the lungs. Follow this artery to the cervical vertebrae and notice that it branches into the dorsal musculature of the back. Note the **vagus** and **phrenic nerves,** which pass next to or over the subclavian artery.

9. At the level of the first rib, each subclavian artery gives off an **internal**

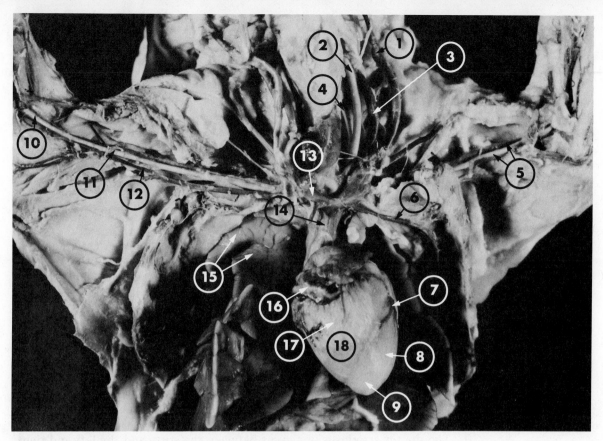

FIGURE 93 *Heart and lungs exposed to show arteries and veins*

1. External Jugular Vein, Left
2. _____
3. Internal Jugular Vein, Left
4. Vagus Nerve
5. Axillary Artery and Vein, Left
6. Internal Thoracic Vein

7. Coronary Artery
8. _____
9. Apex of Heart
10. Brachial Artery, Right
11. Axillary Artery, Right
12. Axillary Vein, Right

13. Brachiocephalic Vein
14. Brachiocephalic Artery
15. _____
16. Atrium, Right
17. Ventricle, Right
18. _____

    mammary artery, which supplies the ventral muscular wall of the fetal pig and of humans.

10. The **costocervical artery** emerges from the subclavian artery at the level of the first rib. This vessel passes dorsally to supply the muscles of the scapular region. A short distance from the costocervical artery is the **thyrocervical trunk.** The thyrocervical supplies the deep muscles of the neck and shoulder region.

11. As the right subclavian artery leaves the thoracic cavity, it is termed the **axillary artery.** As the axillary artery extends into the arm, it is termed the **brachial artery.** Below the elbow, the brachial artery becomes the **radial artery.**

Divisions of the Axillary Artery

12. Return to the right axillary artery and notice the slender **ventral thoracic artery** which emerges from the ventral side of the axillary artery to supply the Pectoralis and Latissimus dorsi muscles.

13. The **subscapular artery** branches from the axillary artery to supply the shoulder or subscapular muscular region. The subscapular artery then branches laterally and posteriorly to give rise to the **lateral thoracic artery,** which supplies the Teres and Latissimus dorsi muscle groups.

Branches of the Aorta in the Thoracic Cavity

14. Return to the aortic arch area and note the descending thoracic aorta in the dorsal region below the aortic arch. In order to observe this, you will need to retract the lungs and heart to one side. Use your probe to

tease and separate the connective tissue that surrounds the aorta in this region.

15. Note the pairs of **intercostal arteries** (10 pairs) branching from the aorta, which supply the intercostal musculature.

16. The **esophageal arteries** have variable origin but travel the length of the esophagus.

VEINS ASSOCIATED WITH THE HEART AND THE THORACIC CAVITY

You may find some variation in the way veins unite with one another. The veins in your specimen may not be as well injected as the arteries. Therefore, some of the veins may not be as easily identifiable as in the previously identified arteries of the thoracic region. Observe other specimens in the laboratory and any display specimens your instructor may have available.

17. The **superior** (*anterior* in pig) **vena cava** is the largest blood vessel entering the right atrium from the superior portion of the body where it collects blood from the head and arms.

18. Use your probe to dissect out the **inferior** (posterior in pig) **vena cava,** which enters the right atrium from the posterior portion of the body.

19. The **pulmonary veins** may best be viewed by retracting the heart laterally and observing these vessels as they enter the left atrium. This group of veins carries blood from the lungs to the heart.

20. The **ductus arteriosus** is a shunt between the aorta and the left and right pulmonary arteries. Thus part of the blood goes from the right ventricle and the remainder to the systemic pattern via the aorta. This vessel stops the blood from completely circulating through the lungs.

Divisions of the Superior Vena Cava

21. The right and left **innominate (brachiocephalic)** veins are the anterior divisions of the superior vena cava. These blood vessels drain blood from the head and arm regions.

22. The **hemiazygos vein** is the first branch of the superior vena cava. Retract the right lung and heart to expose it. Observe this vein on the right side against the vertebral column. The branches of the azygos are the **intercostal** and **esophageal veins.**

23. The **costocervical vein** enters the **innominate vein (brachiocephalic)** at the anterior portion of the superior vena cava. The costocervical vein has one tributary, the **vertebral vein,** which drains the cervical vertebral area.

24. The **internal mammary vein (sternal)** is the second branch entering into the superior vena cava at the level of the third rib. It then passes through the thoracic body wall to supply the mammary glands.

Divisions of the Innominate (Brachiocephalic) Veins

25. The **external jugular veins** are the largest veins in the neck that drain from the cranial region. These vessels are on the ventrolateral surface of the neck just under the platysma muscle. The **internal jugular veins** run parallel to the external jugulars and extend from the base of the skull to the inferior portion of the larynx, where they usually empty into the external jugular vein, although they may enter the innominate vein.

26. The left and right **subclavian veins** are the short proximal segments of the main blood vessel that enters the forelegs. As soon as the subclavian veins pass out of the thoracic cavity they are called **axillary veins.** More distally, in the upper arm region they are known as the **brachial veins,** which are formed by the junction of the radial and ulnar veins in the

forearm. The **subscapular veins** drain the scapular region and enter the axillary veins.

Divisions of the External Jugular Vein

27. Remove excess fascia and musculature that may still be covering the external jugular vein. Observe the **transverse scapular vein** in the shoulder region just under the scapular region.

28. The **thoracic duct** of the lymphatic system enters the left external jugular or subclavian vein behind the peritoneum. It is usually difficult to identify these vessels but sometimes they appear as elongated, beaded vessels. At this point, it would be wise for you to review the arteries and veins in the thoracic cavity before dissecting the blood vessels of the abdominal cavity.

ARTERIES OF THE ABDOMEN

29. The **celiac artery** is the first major branch of the abdominal portion of the descending aorta. Move the viscera laterally and use your blunt probe to remove the peritoneum from the dorsal abdominal wall just below the diaphragm. The celiac artery divides into three main branches: the hepatic, left gastric, and gastrosplenic arteries.

30. The **hepatic artery** passes to the liver and lies parallel to the common hepatic duct and portal vein. The hepatic artery divides into a **gastroduodenal artery**, which supplies the pyloric portion of the stomach, and a **pancreaticoduodenal artery**, which supplies the duodenum and pancreas.

31. The **superior mesenteric artery** emerges from the ventral surface of the descending abdominal aorta about ½ in. posterior to the celiac artery. This large vessel supplies the small intestine and parts of the colon. The superior mesenteric artery branches into the **posterior pancreaticoduodenal artery**, which supplies the pancreas and duodenum, and the **middle colic artery**, which supplies the ileocecal region. This vessel enters the right atrium along with the posterior vena cava.

32. The left **gastric artery** supplies the lesser curvature and ventral surface of the stomach. The lesser omentum and greater omentum must be excised in order to see this vessel.

33. The **adrenolumbar arteries** arise about 1 in. posterior to the superior mesenteric artery on the lateral surface of the abdominal aorta. One branch supplies the suprarenal gland, while the other larger branch continues into the dorsal body musculature.

34. The **renal arteries** arise again about 1 in. posterior to the adrenolumbar arteries on the lateral surface to supply each kidney. Normally, these arteries do not arise at the same level on both sides and frequently branching occurs before they enter the kidney.

35. The **genital (internal spermatic or ovarian) arteries** are a pair of slender arteries that arise posterior to the renal artery. In the male, trace the **internal spermatic artery** from its origin through the inguinal canal and down to the scrotum. In the female, trace this vessel along the dorsal body wall to the ovary and uterine horn. Usually the ovarian artery is easier to locate than the internal spermatic artery.

36. The **inferior mesenteric artery** is a single vessel that arises from the ventral surface of the abdominal aorta to supply the descending colon and rectum. It has two branches: the **left colic artery**, which supplies the descending colon, and the **superior hemorrhoidal artery**, which supplies the proximal portion of the rectum.

37. The **iliolumbar arteries** are paired vessels arising about ½ in. posterior to

the inferior mesenteric artery. These arteries arise just before the bifurcation of the abdominal aorta. These arteries supply the dorsal body wall of the lumbar region.

VEINS OF THE ABDOMEN

Branches of the Inferior Vena Cava

38. You have identified the inferior vena cava, which enters the right atrium from the posterior portion of the body. This vessel proceeds through the diaphragm and lies adjacent to the thoracic portion of the aorta. You may observe many small branches, including the phrenic veins, which are embedded in the dorsal thoracic wall.

39. The **hepatic veins** are three or four veins which may or may not be injected and are best viewed by scraping away some of the liver tissue. Use your blunt probe to expose these vessels in the right medial lobe of the liver.

40. The **adrenolumbar veins** drain the suprarenal glands. The **right adrenolumbar** vein drains into the inferior vena cava and the left drains into the left renal vein.

41. The **renal veins** drain blood from the kidneys. Normally, one vein enters the posterior vena cava more anteriorly than the other. Occasionally you may note two renal veins on the right side.

42. The **genital (spermatic or ovarian)** veins are injected poorly and may not be easily observable. The **right spermatic vein** or **right ovarian vein** drains directly into the inferior vena cava, whereas the left set empties into the left renal vein.

43. The **umbilical vein** passes from the placenta through the umbilical cord to join the hepatic portal vein at the liver. The umbilical vein is joined within the liver to the hepatic vein via the **ductus venosus**. After birth both these veins undergo atrophy and degeneration.

44. The **hepatic portal system** is comprised of branches of the **hepatic portal vein** which drain the stomach, spleen, intestines, and colon. Unless your specimen is triple injected, these branches of the hepatic portal vein will be extremely difficult to observe. Consult Figures 92 and 94 and a display specimen for the interrelationship of the hepatic portal system to the systemic circulation.

45. The **iliolumbar veins** lie adjacent to the iliolumbar arteries against the dorsal body wall. They serve to drain the back. The iliolumbar veins are located just anterior (superior) to the bifurcation of the inferior vena cava into the hind legs.

At this point, it would be wise for you to review the arteries and veins in the abdominal cavity before dissecting the blood vessels of the hindleg.

ARTERIES OF THE HINDLEG

46. Locate the **dorsal abdominal aorta** and note that it gives off many branches into the hindlegs. The dorsal aorta continues as the small **caudal (sacral) artery** into the tail and divides into many branches in the pelvic basin.

47. The descending abdominal aorta bifurcates into the **external iliac arteries,** which are large vessels that pass through the body wall and into the hindlegs. The external iliac artery extends a short distance in the thigh; then a vessel—the **deep femoral artery**—branches medially off the external iliac artery, which continues as the **femoral artery**. The **popliteal**

FIGURE 94 *Pelvic basin exposed to show blood vessels*

1. Kidney, Left

2. Ureter, Left

3. Urinary Bladder, Cut Edge

4. _____

5. Femoral Artery, Right

6. Internal Iliac Artery

7. External Iliac Artery

8. _____

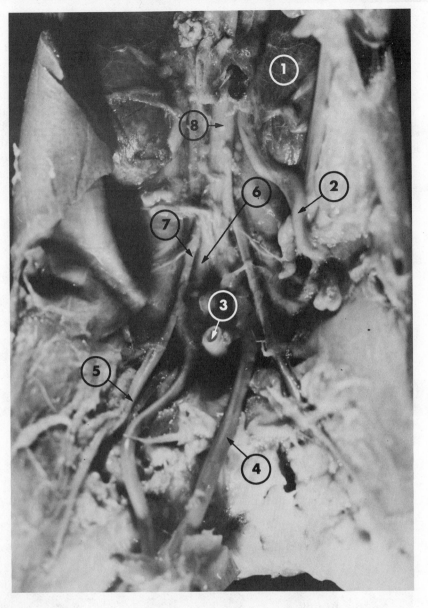

artery branches off the femoral artery and supplies the ventral and lateral portions of the thigh.

48. The **internal iliac artery** is found immediately posterior (inferior) to the origin of the external iliac artery. This artery supplies the uterus and rectum. The internal iliac gives off as its first branch a large **umbilical artery** which passes ventrally out through the abdominal wall within the umbilical cord. After birth the umbilical artery degenerates but the branches remain. The following are tributaries of the umbilical artery but are not often injected: the **superior gluteal artery, hemorrhoidal artery,** and **inferior gluteal artery.** Consult Figures 92 and 95 for this detail.

49. Identify the inferior vena cava in the pelvic basin and note that it bifurcates into a right and left **common iliac vein** (see Figure 95). These veins are closely associated with the iliac arteries and should be found without much teasing of connective tissue. The common iliac vein continues as a large vein into the thigh, the **external iliac vein.** The external iliac continues as the **femoral veins** on the ventral and median surface of the

leg and terminates as the **saphena magna vein** or **greater saphenous vein** on the posterior superficial surface of the calf. The **popliteal vein** may be found entering the femoral vein posterior to the knee draining the posterior region of the thigh.

50. Return to the base of the common iliac veins and observe the **internal iliac veins,** which branch about ¼ in. from the bifurcation of the vena cava into the common iliac veins. The tributaries of the internal iliac vein are deep in the pelvic basin and may be poorly injected. The anterior branch is the **hypogastric vein** or **middle hemorrhoidal vein,** which runs from the urinary bladder. The posterior branch of the internal iliac vein is the **gluteal vein,** which may be found adjacent to the pubic symphysis and rectum.

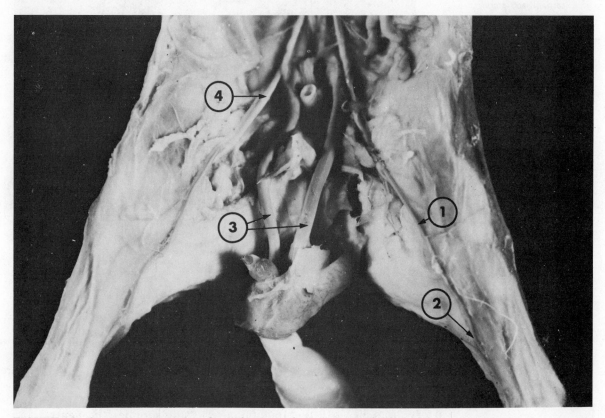

FIGURE 95 *Iliac region exposed to show blood vessels posterior to the kidney*

1. Femoral Vein   3. Umbilical Arteries

2. Popliteal Vein   4. External Iliac Artery

EXERCISE 4   *Dissection of the Pig and Sheep Hearts*

The pig and sheep hearts, like those of all other mammals, consist of four completely separated chambers: two superior atria and two inferior ventricles. Blood from all regions of the body empties into the right atrium by way of the superior (anterior in pig and sheep) and inferior vena cava (posterior in pig and sheep). Blood then passes to the right ventricle. Between the right atrium and right ventricle is the tricuspid valve, which has three flaps or cusps. From the right ventricle, blood is pumped through the pulmonary artery which branches and leads to the right and left lungs where the blood is oxygenated. The oxygenated blood returns from the lungs to the left atrium through the pulmonary veins and then passes into the left ventricle. Between the left atrium and left ventricle is the bicuspid or mitral valve. Attached to the cusps of the bicuspid and tricuspid valves are slender fibers—the **chordae tendineae**—which are attached to the muscular wall of the heart by **papillary muscles**. The tricuspid and bicuspid valves prevent the backward flow of blood from the ventricles into the atria during ventricular systole. The chordae tendineae and the papillary muscles serve to prevent the cusps of the valves from being forced open into the atria during the contraction phase of the heart cycle.

During ventricular systole, blood from the left ventricle flows through the aortic arch. The **aortic semilunar** and **pulmonary semilunar valves** prevent the backflow of blood from the aorta and pulmonary artery into the left and right ventricles, respectively. The thick muscular wall of the heart is the **myocardium**. The **atrioventricular** septum is the thick portion of the myocardium that separates the two atria and ventricles. More specifically, the muscular wall that divides the atria is the **interatrial septum** and the wall that divides the ventricles is the **ventricular septum**.

MATERIALS

preserved and double-injected pig
preserved sheep heart
dissecting instruments and tray

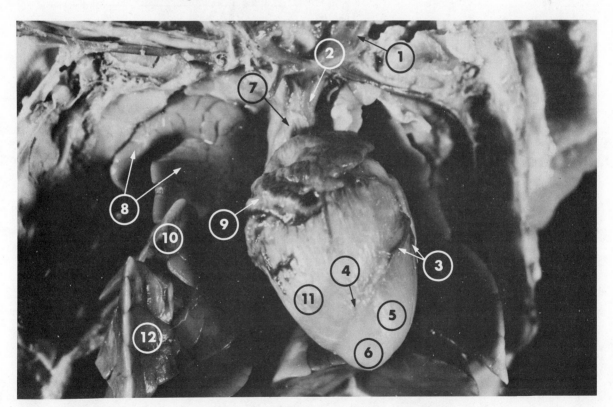

FIGURE 96 *Heart and lungs in situ with pericardium removed, ventral aspect*

1. Left External Jugular Vein	5. Left Ventricle	9. Right Atrium
2. Brachiocephalic Vein	6. Apex	10. Medial Lobe of Right Lung
3. Coronary Arteries	7. Brachiocephalic Artery	11. Right Ventricle
4. Coronary Sulcus	8. Superior Lobe of Right Lung	12. Inferior Lobe of Right Lung

PROCEDURE

DISSECTION OF THE PIG HEART

1. If you have not done so in the preceding exercise, remove any excess fat, the thymus gland, and connective tissue that may obstruct your view of the heart.

2. Identify the following external features of the pig heart (See Figure 96):
   a. **Pericardium**: The pericardium consists of an outer, fibrous **parietal** pericardium (pericardial sac) that surrounds the heart. The inner serous layer is the **visceral pericardium (epicardium),** which lies immediately over the myocardium. The space between the inner and outer

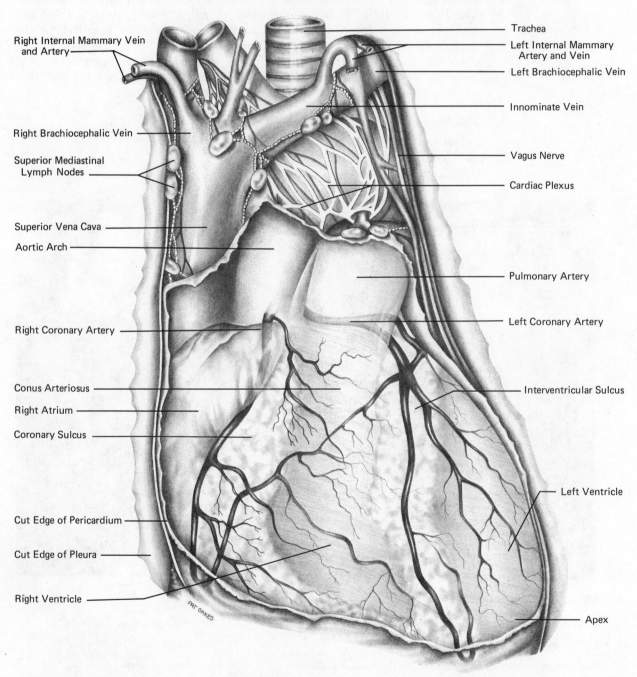

Right Internal Mammary Vein and Artery

Trachea

Left Internal Mammary Artery and Vein

Left Brachiocephalic Vein

Innominate Vein

Right Brachiocephalic Vein

Superior Mediastinal Lymph Nodes

Vagus Nerve

Cardiac Plexus

Superior Vena Cava

Aortic Arch

Pulmonary Artery

Right Coronary Artery

Left Coronary Artery

Conus Arteriosus

Interventricular Sulcus

Right Atrium

Coronary Sulcus

Left Ventricle

Cut Edge of Pericardium

Cut Edge of Pleura

Right Ventricle

Apex

FIGURE 97 *Human heart in situ with pericardial sac removed, ventral aspect*

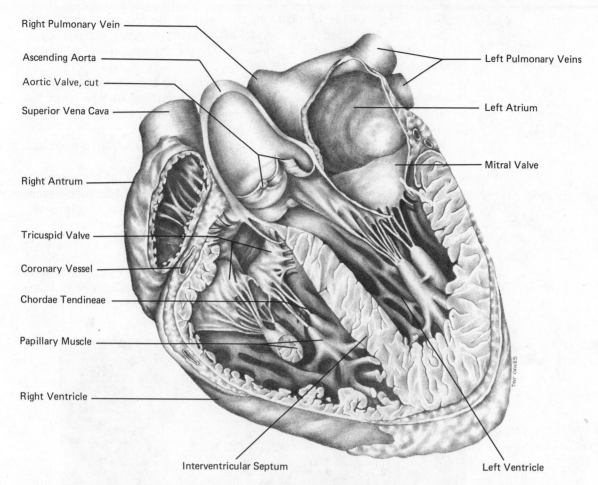

Right Pulmonary Vein

Ascending Aorta

Aortic Valve, cut

Superior Vena Cava

Right Antrum

Tricuspid Valve

Coronary Vessel

Chordae Tendineae

Papillary Muscle

Right Ventricle

Left Pulmonary Veins

Left Atrium

Mitral Valve

Interventricular Septum

Left Ventricle

FIGURE 98 *Human heart showing chamber detail, ventral aspect*

walls of the pericardial sac is the **pericardial cavity.** The pericardial sac will normally not be present on the sheep heart specimen.

b. **Base:** The superior end of the heart where the blood vessels emerge.

c. **Apex:** The inferior end of the heart, which is pointed and in contact with the diaphragm.

d. **Atria:** The atria are two irregular and thin-walled chambers at the base of the heart. The two atria are separated from the ventricles by a horizontal groove, the **coronary sulcus.** The scalloped border of each atrium is the **auricle.**

e. **Ventricles:** These are large muscular posterior chambers of the heart. The left ventricle is more muscularized and therefore firmer than the right ventricle. Externally, the left and right ventricles are separated by a faint groove or depression that extends obliquely across the ventral surface from the base to the apex.

f. **Anterior vena cava** (superior vena cava in man): This vessel empties into the anterodorsal surface of the right atrium.

g. **Posterior vena cava** (inferior vena cava in man): This posterior vessel empties just inferior to the superior vena cava on the anterodorsal surface of the right atrium.

h. **Pulmonary artery:** This vessel emerges from the right ventricle and passes laterally toward the left, where it divides into the right and left pulmonary arteries. After this division, the vessel continues as the **ductus arteriosus** and connects with the aorta. This vessel bypasses the majority of the blood from the lungs to the systemic circulation. Normally, this vessel disintegrates after birth; however, persistence of the opening represents a congenital heart defect known as **patent ductus arteriosus.**

i. **Aortic arch and aorta**: The aortic arch emerges from the left ventricle at a superior angle. In the pig, two major branches emerge from the arch, the brachiocephalic (innominate) artery on the right side and the subclavian artery to the left.

j. **Pulmonary veins**: Small vessels that empty into the left atrium on the anterodorsal surface.

k. **Coronary arteries**: These are blood vessels that supply the myocardium of the heart. These arteries originate at the base of the aortic arch. One coronary artery may be viewed deep between the pulmonary artery and the right auricle.

3. Identify the following internal features of the pig heart after you have removed the heart from the thoracic cavity by cutting the major blood vessels. Using your scalpel, make a frontal incision from the apex to base dividing the heart into dorsal and ventral halves. Wash and probe out any latex that may be present in the chambers (see Figure 98).

FIGURE 99 *Sheep heart in situ with pericardial sac removed, ventral aspect*

1. _____
2. Pulmonary Arteries
3. Pulmonary Artery
4. Atrium, Left
5. _____
6. _____
7. Coronary Sulcus
8. Ventricle, Right
9. Atrium, Right
10. Superior Vena Cava
11. Aorta

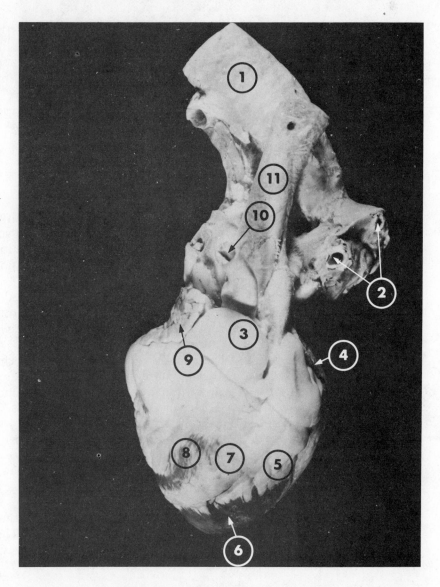

FIGURE 100 *Sheep heart in situ with pericardial sac removed, dorsal aspect*

1. _____

2. Pulmonary Veins

3. Inferior Vena Cava

4. Right Ventricle

5. _____

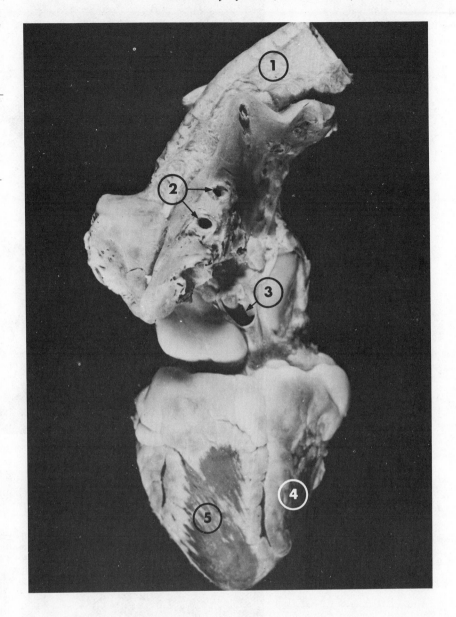

a. **Right and left atria:** The inner surface of these chambers are composed of myocardium. Note the openings in the dorsal wall of the left atrium for the pulmonary veins. In the fetal heart an opening exists between the two atria. Use your probe and identify this hole. This opening is termed the **foramen ovale.** Normally, this closes after birth; however, failure to close results in a defect known as **atrial septal defects.**

b. **Right ventricle:** The inner surface of the ventricular wall is corrugated by muscular cords termed **trabeculae cornae.** Identify the tricuspid valve between the right atrium and right ventricle. Each of the three cusps are connected by thin fibers—the chordae tendineae—and by pillars of **papillary muscles** to the myocardial wall. Note the pulmonary semilunar valves at the base of the pulmonary artery.

c. **Left ventricle:** Identify the **bicuspid valve** or mitral valve between the left atrium and left ventricle. The components of this valve are attached in the same manner as those of the tricuspid valve. Note the aortic semilunar valves, which open into the aortic arch.

FIGURE 101 *Sheep heart with ventral portion reflected to show chamber detail*

1. _____
2. Aorta
3. Pulmonary Arteries
4. Semilunar Valve
5. Mitral Valve (Bicuspid)
6. Chordae Tendineae
7. _____
8. _____
9. Myocardium
10. Interventricular Septum
11. Pericardium
12. Tricuspid Valve

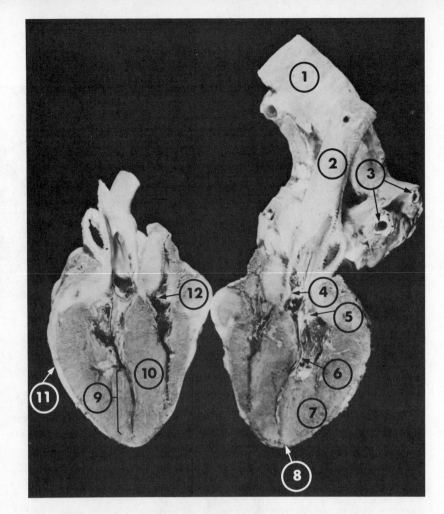

FIGURE 102 *Sheep heart showing chamber and valve detail*

1. Aorta
2. _____
3. Tricuspid Valve
4. _____
5. Interventricular Septum
6. Apex
7. Myocardium of Left Ventricle
8. Aortic Semilunar Valve

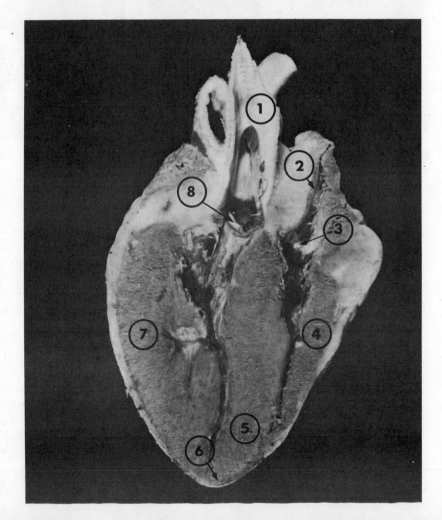

DISSECTION OF THE SHEEP HEART

1. Rinse the sheep heart with cold water in order to remove excess preservative or blood. Make sure the water flows through all of the blood vessels to irrigate any blood clots out of the chambers.
2. Identify the external features of sheep heart from the description given previously under the external features of pig heart (Figures 99 and 100).
3. Identify the internal features of the sheep heart from the description given previously under the internal features of pig heart. The foramen ovale and adult condition ductus arteriosus are not found in sheep or man in internal/external views (see Figures 101 and 102).

B. Blood Physiology

PURPOSE	The purpose of Unit VIII-B is to familiarize the student with basic human blood physiology.

OBJECTIVES	To complete Unit VIII-B the student must be able to do the following:

1. Discuss the significance of coagulation time and the mechanism of normal blood clotting.
2. Demonstrate an understanding of and ability to determine ABO and Rh blood types.
3. Demonstrate an understanding of and ability to determine coagulation times, hemoglobin, and hematocrit percentages.
4. Demonstrate proficiency in blood pressure determination and an understanding of its medical significance.

EXERCISE 1   *Coagulation Time (Clot Formation)*

Coagulation time is the period of time required for a blood sample to form a clot. The clot consists of a mass of tangled strands of **fibrin** in which red and white blood cells are enmeshed. The four essential substances necessary for coagulation to occur are (1) *prothrombin,* which is normally present in the circulatory system and is produced by the liver; (2) *thromboplastin,* which is liberated from injured tissues and platelets; (3) *calcium* (in free ionic form); and (4) *fibrinogen,* a soluble protein present in the plasma. Vitamin K, which is found in green leafy foods and is produced by certain intestinal bacteria, is utilized by the liver in order to produce prothrombin. In the absence of vitamin K or when there is a reduced supply, the liver fails to produce the necessary prothrombin, the result of which is that the blood either fails to clot or clots very slowly. A clot that is formed within a blood vessel and remains attached at the site of formation is termed a **thrombus.** Any clot that has broken loose from its site of formation and is freely floating is termed an **embolus.**

MATERIALS

sterile cotton
unheparinized capillary tubes
70% ethyl alcohol

sterile lancets (Hemolets)
clock with a second hand

PROCEDURE

1. Finger puncture provides the best method of obtaining blood for the tests to follow in the blood studies series. The following finger puncture procedure is to be repeated for obtaining blood samples. Clean the fourth or fifth finger with cotton moistened with 70% ethyl alcohol. Allow the alcohol to evaporate. Remove the sterile Hemolet from the envelope by its blunt end. It is best to hold the Hemolet near its pointed tip and then puncture the finger with a quick jabbing motion. The blood should flow freely of its own accord from the wound. *Do not squeeze the tip of the finger* to obtain blood. If you must squeeze, start at the base of the finger and move toward the tip. If an additional puncture is required, discard the Hemolet used above and obtain a fresh sterile Hemolet for a subsequent puncture.

2. Obtain a drop of blood from the puncture, but do not use the first drop since it usually coagulates fast. After the second drop has formed, place the end of the unheparinized capillary tube in the drop and allow the tube to fill by capillary action. There should be no air bubbles in the column of blood. Repeat the above procedure with another unheparinized capillary tube. Note the time when both tubes were filled and record these times. *Times:*

3. At the *end of 2 minutes,* break off about ¼ inch of the capillary tube and observe whether the blood has clotted, as evidenced by a thread of coagulated blood connecting the two pieces of tubing. Repeat this procedure with both unheparinized capillary tubes at 30-second intervals until the blood has clotted. Record the times. The coagulation times of the two capillary tubes should be approximately the same. Normal coagulation times fall within the range of 2–4 minutes. What explanation can you give for this? Is there any difference in coagulation times between the two tubes?

4. After completion of the exercise, dispose of all equipment.

EXERCISE 2   *ABO and Rh Determination*

Red blood cells contain proteins termed **antigens** (agglutinogens), which are located on the **erythrocyte** (red blood) cell surface. Common antigens found in the human system are A, B, AB (both A and B), O, and D (Rh) antigen. Research has identified more than 100 erythrocyte antigens, each an expression of an inherited gene. To the individual who possesses it, an antigen is beneficial, but this same antigen can be harmful to someone else who did not inherit that particular antigen. When an individual is exposed to *foreign antigens,* this may cause that individual to produce a chemical whose purpose is the destruction of this foreign antigen. This chemical substance, which is contained in the serum fraction of the blood, is called an **antibody**. Once produced, an antibody remains in the circulation for many years, ready to destroy foreign antigens if introduced into the blood. *Antibodies are specific;* they will destroy only an antigen that is identical to the one that stimulated its production and are therefore harmless to other antigens within that same individual.

In order to determine the ABO blood type, two common antibodies, contained in *anti-A,* and *anti-B* anti-sera, will be used. Research indicates that

41% of the United States Caucasian population possess only the A antigen, 9% have only the B antigen, 4% have both A and B antigens, and the remainder of the population, 46%, have neither A nor B antigen. Thus individuals are said to belong to the blood groups A, B, AB, or O, respectively. If the blood type is A, there will be **agglutination** (clumping) with the anti-A serum. If type B is present, agglutination will take place with anti-B serum. If type AB is present, agglutination by both anti-A and anti-B sera will take place. If type O is present, agglutination by neither anti-A nor anti-B sera will take place.

Human erythrocytes are also classified according to the **Rh factor** as either Rh positive or Rh negative, depending on whether the Rh antigen is present. This antigen is more specifically termed the **D-antigen** after the Fisher-Race nomenclature. In order to determine the presence or absence of the Rh antigen, a small whole blood sample is tested with anti-D serum. An individual is Rh+ if agglutination of erythrocytes takes place with the antiserum and Rh– if no agglutination takes place. About 85% of the U.S. Caucasian population is Rh positive and 15% Rh negative.

MATERIALS

70% alcohol and sterile cotton or wipes   anti-A, anti-B, and anti-D antisera
sterile lancets (Hemolets), toothpicks   light warming box
wax marking pencil   microscope slides

PROCEDURE

1. Use a wax pencil to divide a clean microscope slide into two sections or use a prepared slide for blood typing.
2. Label one side of the slide "A," the other side, "B." Label a second slide "D."
3. Puncture your finger, wipe away the first drop of blood and place the second and third drops on the slide, one on side A, the other on side B. Place another drop of blood on the second slide.
4. Add one drop of anti-A serum to the drop of blood marked A and one drop of anti-B serum to the B side. Place one drop of anti-D antiserum on the other drop of blood, which is on the second slide.
5. Mix the blood and antisera by using a toothpick. In order to avoid contamination, use a separate toothpick for each mixing.
6. Let stand for 2 minutes. Examine closely for agglutination *during this 2-minute* period. If doubtful, use low power of the microscope.
7. Place the microscope slide marked D on the warming box (40°C) and mix by rocking the slide back and forth for *2 minutes*. Examine for agglutination during this time period.
8. Determine your ABO and Rh types. Record your results in the tables provided. Mark agglutination as (+) and no agglutination as (–).

Type	Anti-A	Anti-B	Number of individuals	Class percentage	Theoretical percentage
A					41
B					9
AB					4
O					46

D Antigen	Number of individuals	Class percentage	Theoretical percentage
Rh +			85
Rh −			15

EXERCISE 3   *Hemoglobin Estimation by the Tallquist Method*

The amount of *hemoglobin* present in red blood cells is a good indicator of the oxygen-carrying capacity of the blood. A simple method of measuring hemoglobin is by comparing a small piece of Tallquist paper that has been saturated with a sample of blood with a Tallquist color chart.

MATERIALS

Tallquist scale and test paper
70% ethyl alcohol
sterile lancets and cotton

PROCEDURE

1. Remove one square of paper from the Tallquist booklet.
2. Puncture the finger as previously described. Place the second drop of blood in the center of the paper. Wait *15 seconds* and compare your sample with the Tallquist scale.
3. Record the scale estimate in grams of hemoglobin per 100 cc of blood. Note differences in male and female comparisons and any anemic conditions that may be indicated.

EXERCISE 4   *Hematocrit or Packed Cell Volume*

The percentage of erythrocytes found in a set volume of blood is known as the *hematocrit* or packed cell volume (PCV). The PCV is a reliable indicator of the percentage of red blood cells in a given sample. Centrifugation of the blood sample results in the bottom portion of the tube being packed with red blood cells. The upper portion will contain the plasma fraction of the blood. A hematocrit reader is used to determine the percent of packed red blood cells in the sample. The normal hematocrit in males is 40%–54% with 47% as the average. In females, it is 37%–47%, with 42% the average.

MATERIALS

Clay–Adams microhematocrit reader
  (or substitute)
70% alcohol and sterile cotton
heparinized (0.5 mm) capillary tubes

Clay–Adams Seal-Ease
microhematocrit centrifuge
sterile lancets (Hemolets)

PROCEDURE

1. Puncture your finger as previously described. After the blood flows freely, wipe away the first drop of blood and place the red end of the capillary tube on the wound area and allow the blood to flow downward into the capillary tube. *Fill the tube only with blood* (no bubbles).
2. Place your finger over the opposite end of the capillary tube and place the red end of the tube into the Seal-Ease.

3. Place the sample tube into the microhematocrit centrifuge *with an empty capillary tube opposite the sample tube for balance;* otherwise the sample in the capillary tube will spill into the head of the centrifuge.
4. After loading the centrifuge, tighten the inside and outside covers. Set the timer for 4 minutes.
5. After the 4-minute spin period, remove the sample and place in the microhematocrit reader. Determine the percentage of blood volume from the scale.
6. Record your results.

EXERCISE 5   *Blood Pressure and Pulse Determinations*

The beating heart makes certain characteristic sounds that can be heard through an instrument called a *stethoscope.* The stethoscope amplifies well-defined sounds, described as lubb–dubb. The first sound (lubb), which occurs during ventricular **systole**, takes place during the contraction of the ventricles and is caused by the closing of the atrioventricular valves. The second sound (dubb) is produced during ventricular **diastole** and is the sound produced by the closing of the semilunar valves. The sounds heard between these two may indicate the condition of the valves of the heart. During ventricular systole, a wave of pressure, termed the **pulse**, is produced in the arteries caused by ventricular contraction. The pulse can normally be felt where an artery is near the surface of the skin and over the surface of a bone. Pulse rates may vary in individuals because of changes in temperature, stress factors, the time of day, and other factors.

a. *Taking the Radial Pulse*
Use your index and middle fingers to palpate the pulse. *Do not use the thumb* since a clear pulse is present in the thumb itself. Have your partner place his forearm with the palm up on the laboratory table. Reach over the wrist, not under it. The pulse can be felt behind your partner's thumb between the large tendon and bony prominence found on the lateral aspect of the wrist. Too much pressure will obliterate the pulse. Practice so you can achieve an accurate pulse rate by counting the beats per 15 seconds and multiplying by 4 to obtain the beats per minute. The normal adult range is 72–80 beats per minute. Concentrate fully on the pulse and screen out other noises.

b. *Determination of the Carotid Pulse*
Place your fingers on either side of your partner's larynx (Adam's apple). Gently press medially and toward the back. Apply pressure gradually until you feel the pulse. Relocate your fingers slightly upward or downward until the pulse is clearly felt with at least two of the fingers. Practice several times in order to achieve an accurate pulse rate.

c. *Determination of the Arterial Blood Pressure by Ausculation*
Have your partner extend his forearm across a table with the palm up. The upper arm should be approximately at heart level. Either arm may be used, but the left is preferable. Use the same arm for all exercises involving blood pressure determinations.

Check the **sphygmomanometer** cuff for markings indicating the area to be applied against the brachial artery. Some cuffs have markings for both the right or left arm and some have no such markings. In any case, apply the center of the inflatable area of the cuff on the inside of the upper arm and wrap the long end of the cuff snugly about the limb. If it is not snug, inflate the sphygmomanometer until it does not slip down under the influence of gravity. Place the gauge or mercury column where it can be easily read. Put the earpieces of the stethoscope in the ears. Place the bell or diaphragm of the stethoscope over the blood vessels visible under the skin covering the inside of the elbow.

Inflate the cuff to 180 mm mercury (please, no higher; work quickly). Slightly open the valve at the bulb and listen carefully with the stethoscope as the pressure in the cuff decreases. The point at which the pressure sounds or sounds of Korotkoff are first heard is the **systolic** pressure. The point at which the *pressure sounds* assume a muffled quality represents the *diastolic* pressure. If you cannot detect the change to the muffled sound, then consider the point at which all pressure sounds disappear to be the diastolic pressure. *However you must consistently use one method or the other.* Consult the following table for Normal Blood Pressure Values For Men and Woman. Repeat three times and determine pressure. Wait 15–30 seconds before repeating each determination.

BLOOD PRESSURE VALUES FOR MEN AND WOMEN

Age	Systolic			Diastolic		
	Normal range	Mean	Hypertension lower limit	Normal range	Mean	Hypertension lower limit
Men						
16	105–135	118	145	60–86	73	90
17	105–135	121	145	60–86	74	90
18	105–135	120	145	60–86	74	90
19	105–140	122	150	60–88	75	95
20–24	105–140	123	150	62–88	76	95
25–29	108–140	125	150	65–90	78	96
30–34	110–145	126	155	68–92	79	98
35–39	110–145	127	160	68–92	80	100
40–44	110–150	129	165	70–94	81	100
45–49	110–155	130	170	70–96	82	104
50–54	115–160	135	175	70–98	83	106
55–59	115–165	138	180	70–98	84	108
60–64	115–170	142	190	70–100	85	110
Women						
16	100–130	116	140	60–85	72	90
17	100–130	116	140	60–85	72	90
18	100–130	116	140	60–85	72	90
19	100–130	115	140	60–85	71	90
20–24	100–130	116	140	60–85	72	90
25–29	102–130	117	140	60–86	74	92
30–34	102–135	120	145	60–88	75	95
35–39	105–140	124	150	65–90	78	98
40–44	105–150	127	165	65–92	80	100
45–49	105–151	131	175	65–96	82	105
50–54	110–165	137	180	70–100	84	108
55–59	110–170	139	185	70–100	84	108
60–64	115–175	144	190	70–100	85	110

Reprinted with permission from Eli Lilly and Company, Indianapolis, Indiana.

C. Cardiac Muscle Physiology

PURPOSE	The purpose of Unit VIII-C is to observe the action of the heart and to demonstrate some of the properties of cardiac muscle when stimulated by various agents.

OBJECTIVES   To complete Unit VIII-C, the student must be able to do the following:

1. Observe the normal cardiac cycle of the frog heart by recording a tracing of the frog's heartbeat.
2. Determine the effect of temperature change on the rate of heartbeat and the significance of that effect.
3. Observe the refractory period, extrasystole, and compensatory pause in the cardiac cycle.
4. Observe the effect of vagal nerve inhibition on the cardiac cycle.
5. Determine the effect of adrenalin on the cardiac cycle and its influence on the cardiovascular system.

The frog heart is similar in action to the human heart even though it is a three-chambered heart. The heartbeat is influenced by internal and external factors. Special conducting tissues of the heart are located in small masses of tissue termed the *sinoatrial* (SA) and *atrioventricular nodes* (AV). These specialized centers establish a particular rhythm, which may be accelerated or slowed by specific nerve fibers of the central nervous system. The medulla oblongata and other chemical and physical factors influence the heart rate. In these exercises, you will study the nature of the heartbeat and some factors that influence its rate.

EXERCISE 1   *Observation of the Normal Cardiac Cycle of the Frog*

MATERIALS

dissecting instruments	frog board or dissecting pan
dissecting pins	Ringer's solution
mechanical or electronic recording device and related apparatus	living frog(s)

PROCEDURE

1. Your instructor will explain the use of the recording equipment that is to be used in this exercise.
2. Place a double-pithed frog [see Unit VI-B ("The Physiology of Muscle Contraction") for pithing procedure], ventral surface up, on a frog board or dissecting pan. Pin the feet and lower jaw to the pan.
3. Use your scissors to make a midventral incision through the skin from the pubic symphysis to the midline of the mandible. Cut laterally on each side at the pectoral and pelvic girdles and pin back the flaps of skin. Cut superiorly through the ventral musculature from the pubic symphysis. Carefully cut through the sternum area. Pin back the cut edges of the pectoral girdle in order to expose the heart, lungs, and major blood vessels. Identify all of the external structures of the heart before proceeding to recording of the cardiac cycle.

4. Connect the mechanical or electronic recording device according to your instructor's directions.
5. Record the tracing of the frog's cardiac cycle for 5 minutes. You may moisten the heart with several drops of Ringer's solution during this 5-minute period. The average rate for a frog is between 50–70 beats per minute.
6. Determine the rate of the atria and the ventricle for 15 seconds by counting the number of beats occurring during 15 seconds. Multiply each figure by 4 to determine the beats per minute. Observe the sequence of contraction of the chambers of the frog heart during this 5-minute recording period.

FIGURE A *Normal Frog Cardiac Cycle*

a–b auricular systole
b–c auricular diastole (complete at f)
c–e ventricular systole

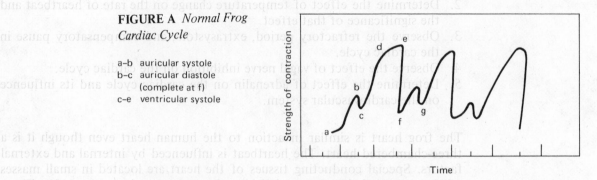

7. Refer to Figure A for a tracing of the *normal cardiac cycle* of the frog. Attach a labeled recording of the normal cardiac cycle of the frog's heart. *Note:* Keep your frog preparation moist and save for the next exercise, which *must* follow immediately.

EXERCISE 2 *Observation of the Effect of Temperature on the Heartbeat*

The heartbeat can be influenced by various factors such as activity, age, body weight, and body temperature. The average heartbeat rate for a frog is between 50 and 70 beats per minute. In this exercise, you will note how temperature change can influence rate of heartbeat. All biological reactions increase with temperature within physiological limits (2°C–40°C). An increase within limits in temperature speeds up the rate of biological reactions. Biologists often use the change of rate of a reaction after a 10° temperature change (Q_{10}) as a measurement of thermal energy:

$$Q_{10} = \frac{\text{rate at a higher temperature}}{\text{rate at a 10° lower temperature}}.$$

MATERIALS
double-pithed frog—same frog as used in previous exercise
8°C and 28°C water baths
thermometer

mechanical or electronic recording device
wax pan

PROCEDURE
1. Use the same frog preparation as in the previous experiment.
2. Immerse the frog in the 8°C water bath. This should be done slowly by adjusting the temperature of the water bath through the addition of Ringer ice and by heating. After the frog has been immersed at 8°C for

15 minutes, remove the frog and record the cardiac cycle with a mechanical or electronic recording device.

3. Repeat the immersion in water of 18°C and 28°C. Calculate the Q_{10} and record your results. Record the cardiac cycle and note changes in the rate and amplitude of contraction.

4. Compute the Q_{10} for the 8°C, 18°C, and 28°C temperatures.

Temperature	Q_{10}
8°C	
18°C	
28°C	

5. Attach a labeled recording of the cardiac cycle of the frog for the above temperatures to Question 10 of the Discussion.

6. Same frog for next exercise. Moisten with Ringer's solution.

EXERCISE 3   *Observation of Refractory Period, Extrasystole, and Compensatory Pause in the Cardiac Cycle*

The refractory period, during which stimuli have no effect, is longer for cardiac muscle than for striated or skeletal muscle. This extended refractory period is an important fact in setting the rhythmicity and in preventing tetany of the heart. If an extra stimulus is applied to the ventricle during systole or diastole, the ventricle will respond to this stimulus by giving an extra stroke or contraction (extrasystole), after which the ventricle may pause and "skip a beat" (compensatory pause). What takes place is that the refractory period to stimuli from the atria causes a longer delay, which results in a compensatory pause of the ventricle.

MATERIALS

pithed frog
mechanical or electronic recording device
electronic stimulator and event–time marker
dissecting instruments
wax pan or frog board

PROCEDURE

1. Use the same frog preparation as in the previous exercise. Keep moist with Ringer's solution.

2. Position the mechanical or electronic recording apparatus. Record a tracing of 8–10 normal beats of the heart.

3. Using an electrode holder, position the electrodes so that the wire tips are in contact with the heart just slightly below the atrial–ventricular groove. Use a single stimulus of 40 volts, stimulate the ventricle *for 2 seconds,* and then break the circuit. *Wait 10 seconds* before applying the next stimulus. Stimulate the ventricle during the systolic and diastolic phases of the cardiac cycle. Continue stimulating until the tracing shows three extra systoles. Note any compensatory pauses that occur between stimuli (see Figure B).

FIGURE B *Refractory Period Tracing*

a refractory period
b–c extrasystole
d compensatory pause

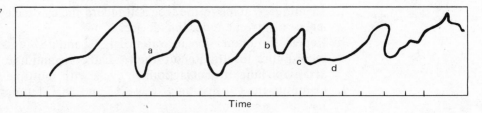

Time

4. Observe the following during the recording period:

 a. The point at which the ventricle exhibits absolute refractoriness:

  _____

 b. The length of the refractory period: _____

 c. The duration of a compensatory pause: _____

5. Attach to Question 10 of the Discussion a labeled recording of the cardiac cycle that exhibits extra systole and compensatory pause.
6. Save the frog for Exercise 4. *Keep moist with Ringer's solution.*

EXERCISE 4

Observation of the Effect of Vagal Nerve Inhibition and Adrenalin Acceleration

The inherent rhythm of cardiac muscle may be modified by the central nervous system as well as by thermal effects. There are two branches of the autonomic nervous system that innervate the heart: the *vagus nerve* by way of the parasympathetic branch and the *nerve fibers* from the sympathetic chain of **ganglia** in the cervical region. When the vagus nerve is stimulated, the cardiac rate may decrease or cease beating for a few moments. This is referred to as vagal inhibition. The secretion of the vagus nerve is acetylcholine, which momentarily slows the heart rate; however, this inhibition is only temporary since acetylcholine is destroyed within 0.05 seconds by the enzyme acetylcholine esterase. The sympathetic fibers secrete **norepinephrine**, which accelerates the heart rate. This effect is also short in duration since enzymatic breakdown also occurs. **Adrenalin (epinephrine)** released from the adrenal medulla during physical or emotional stress has an action similar to that of norepinephrine.

MATERIALS

pithed frog
dissecting instruments
wax pan or frog board
epinephrine (1:1000) or acetylcholine (1:1000)

mechanical or electronic recording
 device
electronic stimulator and event–time
 marker
26-gauge hypodermic needle

PROCEDURE

1. Use the same pithed frog preparation as in the previous exercise. If the condition of the frog is poor, obtain a fresh specimen, pith it, and make a similar preparation.
2. Locate and isolate the vagus nerve, which runs along the external jugular vein. After identifying the nerve, loosely tie a loop of thread around it. A dissecting microscope may be used to assist in isolating this nerve.

3. Using an electrode holder, position the electrodes so the wire tips are in contact with only the nerve.
4. Record a normal cardiac cycle for 15 seconds. Stimulate the vagus nerve with multiple stimuli of 200 volts for a period of 5 seconds. Repeat after obtaining a normal cardiac cycle with multiple stimuli of 10-, 15-, and 20-second durations. From your tracing, determine and label where the following occur: point of vagal inhibition and termination of vagal inhibition (refer to Figure C). Attach the tracing to Question 10 of the discussion page.

FIGURE C *Vagal Inhibition*

a point of vagal stimulation
b vagal inhibition
c recovery

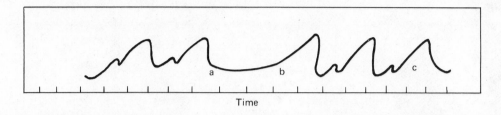

Time

5. Record a normal cardiac tracing for 15 seconds before injection of epinephrine. Inject 0.2 cc of 1:1000 epinephrine intraperitoneally or intramuscularly. *Record for 10 minutes* and note the height of the ventricular systole. Determine the effects on the heart rate and note any other visible bodily changes (see Figure D). Attach the tracing to Question 10 of the Discussion page.

FIGURE D *Epinephrine Stimulation*

a normal cardiac systole and diastole
b point of epinephrine stimulation

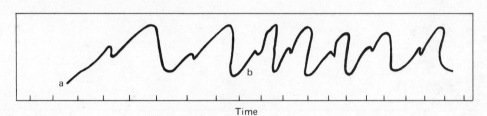

Time

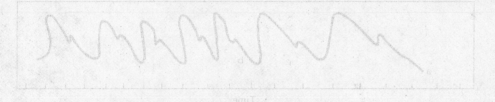

Circulatory System

DISCUSSION

1. Using a flow sheet, describe the scheme of human blood clotting.

_____

_____

_____

_____

_____

2. Where is hemoglobin produced and what is its purpose? Can you suggest reasons why hemoglobin values differ in males and females?

3. Identify the place of formation and function each of the following performs in the coagulation of blood:

    a. Prothrombin: _____

    _____

    b. Fibrinogen: _____

    _____

    c. Thromboplastin: _____

4. Briefly describe how you may identify the following formed elements in a human blood smear:

a. Acidophil: _____

_____

b. Basophil: _____

_____

c. Neutrophil: _____

_____

d. Lymphocyte: _____

_____

e. Monocyte: _____

_____

f. Erythrocyte: _____

_____

g. Platelet: _____

_____

5. Trace the path of a drop of blood from the rectum to the heart and back to the rectum, naming all the blood vessels and structures of the heart through which the blood will pass.

6. Match the following statements with the following structures of the heart:

aorta   inferior vena cava   right atrium
bicuspid valve   left atrium   right ventricle
chordae tendineae   left ventricle   septum
coronary arteries   myocardium   tricuspid valve
endocardium   pulmonary artery   aortic semilunar valve
pericardium   pulmonary vein

a. Chamber that receives blood from lungs: _left atrium_

b. The valve between the left atrium and the left ventricle: _Bicuspid valve_

c. Muscle tissue that composes the heart: _myocardium_

d. Tissue that lines the heart: _Endocardium_

e. Covering of serous tissue that surrounds the heart: _pericardium_

f. Heart's own blood vessels: _Coronary arteries_

g. Left ventricle exit valve: _aortic semilunar valve_

h. Chamber that forces blood into systemic circulation: _____

i. Ventricle partition: _Septum_

7. Match the following statements with the following terms:

anemia   leukocytosis   monocyte
antibodies   lymphocytes
erythrocytes   plasma
fibrin   platelets
lymph   serum

a. Liquid portion of the blood: _Plasma_

b. Phagocytic cells: _monocyte_

c. Cells that transport oxygen: _erythrocytes_

d. Noncellular irregular elements essential to clotting: _platelets_

e. Protective immunity: _antibodies_

f. Low hemoglobin level: _anemia_

g. Too many white blood cells: _leukocytosis_

8. Identify the major organs served by each of the following vessels:

a. Renal artery: _____

b. Splenic artery: _____

c. Inferior mesenteric artery: _____

d. Genital artery: _____

e. Internal mammary artery: _____

f. Hepatic portal vein: _____

g. Celiac artery: _____

h. Adrenolumbar artery: _____

9. Briefly describe the effect the vagus nerve and epinephrine have in regulating the cardiac cycle.

10. Attach and interpret your labeled physiological recordings to this sheet.

11. Using a flow diagram, trace fetal circulation.

UNIT IX *Respiratory System*

PURPOSE	The purpose of Unit IX is to familiarize the student with the anatomy of the human and fetal pig respiratory organs and to allow the student to understand the use of the spirometer and pneumograph in measuring standard respiratory volumes.

OBJECTIVES

To complete Unit IX, the student must be able to do the following:

1. Identify major organs of the human and pig respiratory systems.
2. Identify the respiratory organs and related circulatory structures of sheep pluck.
3. Demonstrate knowledge of basic lung and trachea histology.
4. Determine standard respiratory air volumes using a spirometer.
5. Use a pneumograph to determine respiratory variations.

MATERIALS

embalmed fetal pig
sheep pluck
physiological recording equipment
spirometers and replaceable
   mouthpieces

dissecting pan
dissecting pins
prepared slides of lung and
   tracheal tissue
pneumograph and recording
   attachments

PROCEDURE

EXERCISE 1 *Anatomy of Human and Pig Respiratory Organs*

The function of the respiratory system is to transport oxygen from the external atmosphere to the alveoli of the lungs. The blood then transports oxygen to the cells for metabolic use. The respiratory system also eliminates excessive carbon dioxide from the body. Anatomically the lungs function as the center of gas exchange, but also included as accessory structures are the *nasal cavity* and surrounding **paranasal sinuses**; frontal, sphenoid, ethmoid, and maxillary bones; and the *pharynx, larynx, trachea, bronchi,* and divisions of the bronchi (see Figures 103 and 104 for the structures of the human respiratory system). Notice the number of *lobes* in the human right and left lung. Also note the C-shaped cartilaginous rings of the trachea, which prevent collapse during inspiration (see Figures 105 and 106).

FIGURE 103 *Sagittal section of human head*

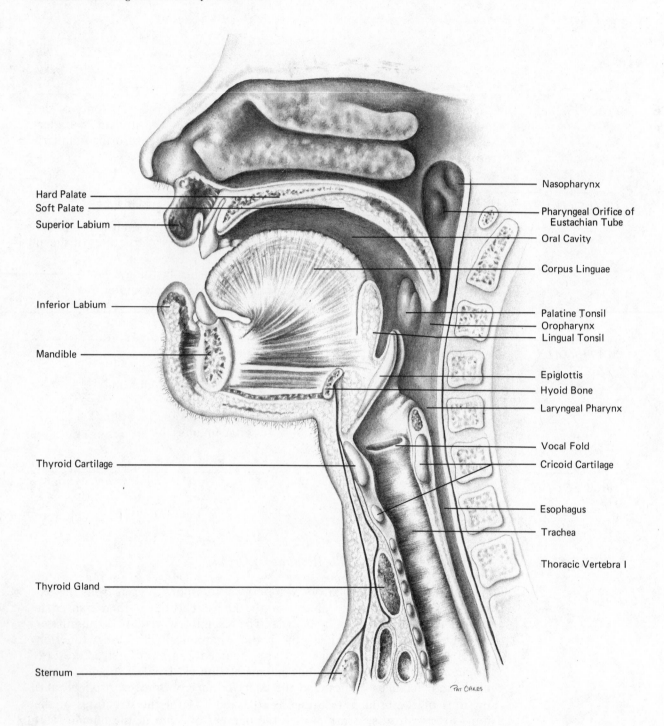

Hard Palate

Soft Palate

Superior Labium

Inferior Labium

Mandible

Thyroid Cartilage

Thyroid Gland

Sternum

Nasopharynx

Pharyngeal Orifice of
Eustachian Tube

Oral Cavity

Corpus Linguae

Palatine Tonsil

Oropharynx

Lingual Tonsil

Epiglottis

Hyoid Bone

Laryngeal Pharynx

Vocal Fold

Cricoid Cartilage

Esophagus

Trachea

Thoracic Vertebra I

FIGURE 104 *Human lung showing termination of bronchiole into alveolus*

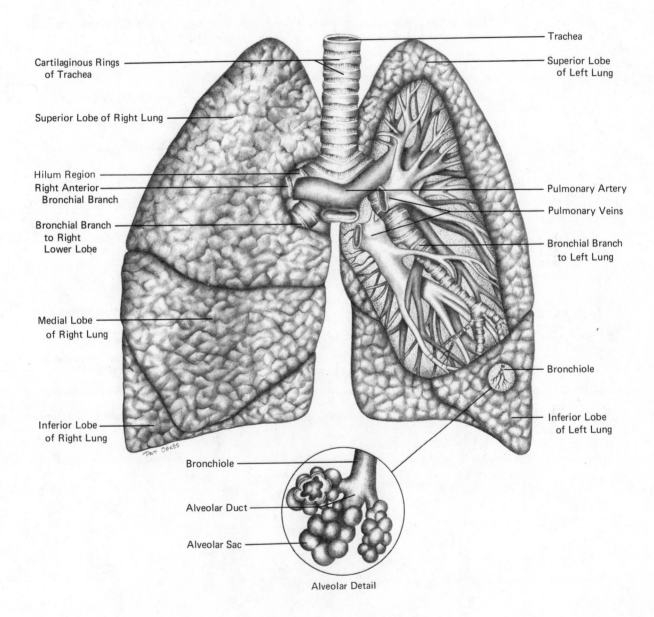

Cartilaginous Rings
of Trachea

Superior Lobe of Right Lung

Hilum Region
Right Anterior
Bronchial Branch

Bronchial Branch
to Right
Lower Lobe

Medial Lobe
of Right Lung

Inferior Lobe
of Right Lung

Trachea

Superior Lobe
of Left Lung

Pulmonary Artery

Pulmonary Veins

Bronchial Branch
to Left Lung

Bronchiole

Inferior Lobe
of Left Lung

Bronchiole

Alveolar Duct

Alveolar Sac

Alveolar Detail

FIGURE 105   *Human larynx viewed from dorsal and superior aspects* (Redrawn from B.J. Anson and C.B. McVay: *Surgical Anatomy,* Philadelphia: W.B. Saunders, Co., 1971)

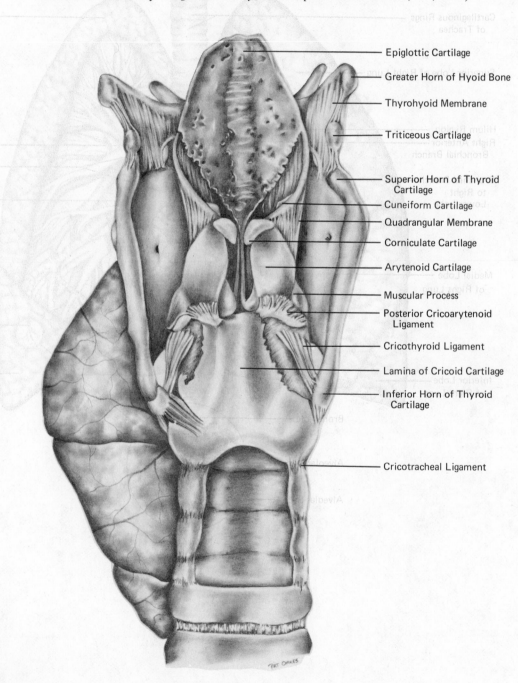

Epiglottic Cartilage

Greater Horn of Hyoid Bone

Thyrohyoid Membrane

Triticeous Cartilage

Superior Horn of Thyroid Cartilage

Cuneiform Cartilage

Quadrangular Membrane

Corniculate Cartilage

Arytenoid Cartilage

Muscular Process

Posterior Cricoarytenoid Ligament

Cricothyroid Ligament

Lamina of Cricoid Cartilage

Inferior Horn of Thyroid Cartilage

Cricotracheal Ligament

FIGURE 106 *Human larynx viewed from lateral and superior aspects* (Redrawn from B.J. Anson and C.B. McVay: *Surgical Anatomy,* Philadelphia: W.B. Saunders, Co., 1971)

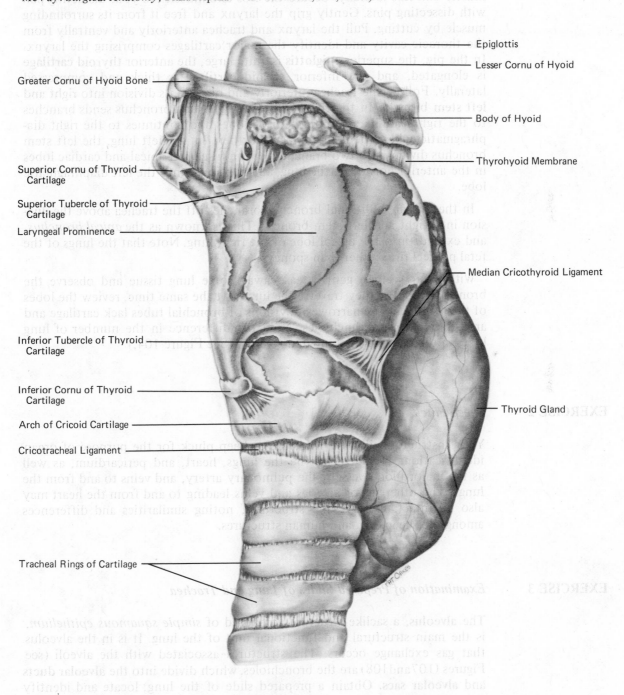

Epiglottis

Lesser Cornu of Hyoid

Greater Cornu of Hyoid Bone

Body of Hyoid

Thyrohyoid Membrane

Superior Cornu of Thyroid Cartilage

Superior Tubercle of Thyroid Cartilage

Laryngeal Prominence

Median Cricothyroid Ligament

Inferior Tubercle of Thyroid Cartilage

Inferior Cornu of Thyroid Cartilage

Thyroid Gland

Arch of Cricoid Cartilage

Cricotracheal Ligament

Tracheal Rings of Cartilage

The nostrils of the fetal pig are small and are located on the anterior surface of the snout, or **rostrum**. The nasal cavity is long and narrow. If you have not already done so, continue the midventral incision you made when studying the digestive and circulatory systems so as to expose the pharynx, or throat, and larynx, which is a cartilaginous, elongated, boxlike structure at the superior end of the trachea. The **trachea**, or "windpipe," can be recognized by its cartilaginous rings.

Expose the respiratory structures in your fetal pig by retracting the cut muscle and ribs laterally. Secure the ribs and muscles to your dissecting pan with dissecting pins. Gently grip the larynx and free it from its surrounding muscle by cutting. Pull the larynx and trachea anteriorly and ventrally from the thoracic cavity and identify the major cartilages comprising the larynx. In the pig, the superior **epiglottis** is quite large, the anterior **thyroid cartilage** is elongated, and the inferior **cricoid cartilage** is thick and compressed laterally. Follow the trachea inferiorly and observe its division into **right and left stem bronchi.** In the right lung, the right stem bronchus sends branches to the right cardiac and intermediate lobes and continues to the right diaphragmatic lobe, where it branches further. In the left lung, the left stem bronchus divides into two branches for the fused left apical and cardiac lobes in the anterior portion of the lung, and continues into the left diaphragmatic lobe.

In the pig, an additional bronchus branches off the trachea above the division into right and left stem bronchi. This is known as the *apical bronchus,* and extends into the apical lobe of the right lung. Note that the lungs of the fetal pig feel firm rather than spongy.

With your scalpel, gently tease away some lung tissue and observe the bronchial tubes as they traverse the lungs. At the same time, review the lobes of the pig lungs. The narrowest divisions of bronchial tubes lack cartilage and are referred to as **bronchioles.** Note the difference in the number of lung lobes in the fetal pig as compared to man. (See Figure 104.)

EXERCISE 2   *Sheep Pluck*

Your instructor will have available a sheep pluck for the purpose of organ identification. Included will be the lungs, heart, and pericardium, as well as the major blood vessels, the pulmonary artery, and veins to and from the lungs. The other major arteries and veins leading to and from the heart may also be intact. Observe the structures, noting similarities and differences among the sheep, pig, and human structures.

EXERCISE 3   *Examination of Prepared Slides of Lung and Trachea*

The **alveolus,** a saclike structure composed of *simple squamous epithelium,* is the main structural and functional unit of the lung. It is in the alveolus that gas exchange occurs. The structures associated with the alveoli (see Figures (107 and 108) are the bronchioles, which divide into the **alveolar ducts** and **alveolar sacs.** Obtain a prepared slide of the lung; locate and identify the previously named structures. Also obtain a prepared slide of the trachea

FIGURE 107 *Alveolar detail of lung*

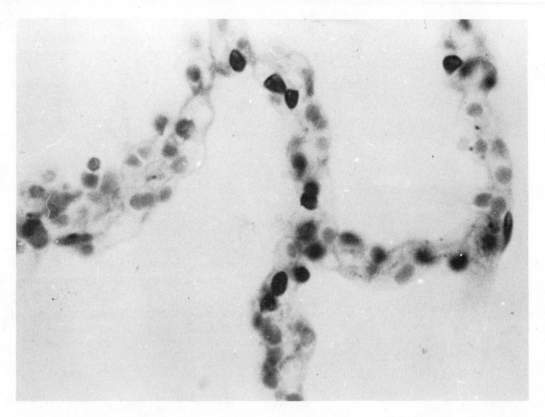

FIGURE 108 *Detail of bronchus in lung*

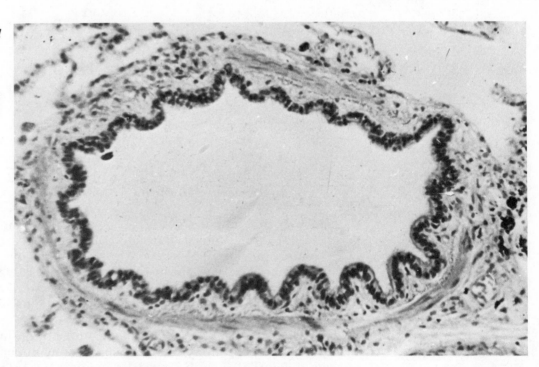

(cross section) and observe the tissue types present. Note the ciliated pseudo-stratified epithelium, hyaline cartilage, and trachealis muscle. *Sketch and label* the structures that you observe.

Lung:

Trachea:

EXERCISE 4   *Spirometric Measurement of Standard Respiratory Volumes*

Variations in size, age, and sex of an individual account for variations in respiratory volumes. Normal breathing usually moves about 500 ml of air in and out of the lungs and is referred to as *tidal volume* (TV). Even after a normal inspiration, an individual can forcibly inhale approximately 3000 ml of air into the lungs. This forced inspiration is called the *inspiratory reserve volume* (IRV). Also after a normal expiration, an individual can forcibly exhale approximately 1100 ml of air or the *expiratory reserve volume* (ERV). The *vital capacity* is the sum total of the inspiratory reserve, the tidal, and the expiratory reserve volumes, or approximately 4600 ml of air; thus the vital capacity is the total exchangeable air of the lungs. In addition, approximately 1200 ml of air remain in the conducting tubules and this air is not exchangeable. This air is referred to as **residual volume** (RV).

Air volumes can be measured by means of a spirometer. Set the needle of the spirometer at zero (0) at the beginning of each use. Remember each student should use a clean replaceable mouthpiece. Each student should determine his *tidal volume, expiratory reserve volume, vital capacity,* and *inspiratory reserve volume* (vital capacity minus expiratory reserve and tidal volumes). See the table below to determine normal ranges for vital capacity. Enter individual volumes on the discussion page (question 1) of this unit

NORMAL VITAL CAPACITY (IN CUBIC CENTIMETERS) OF ADULT MALES AND FEMALES*

Height in inches	Age in years:	20	30	40	50	60	70
Males:	60	3885	3665	3445	3225	3005	2785
	62	4154	3925	3705	3485	3265	3045
	64	4410	4190	3970	3750	3530	3310
	66	4675	4455	4235	4015	3795	3575
	68	4940	4720	4500	4280	4060	3840
	70	5206	4986	4766	4546	4326	4106
	72	5471	5251	5031	4811	4591	4371
	74	5736	5516	5516	5076	4856	4636
Females:	58	2989	2809	2629	2449	2269	2089
	60	3198	3018	2838	2658	2478	2298
	62	3403	3223	3043	2863	2683	2503
	64	3612	3432	3252	3072	2892	2710
	66	3822	3642	3462	3282	3102	2922
	68	4031	3851	3671	3491	3311	3131
	70	4270	4090	3910	3730	3550	3370
	72	4449	4269	4089	3909	3729	3549

*Variations must be at least 20% below predicted normal to be considered subnormal. Variations can also exist depending upon size and body structure.
Adapted with permission from Propper Manufacturing Co., Inc., New York.

EXERCISE 5 *Use of the Pneumograph in Determination of Respiratory Variations*

Human respiratory variations are determined by use of a *pneumograph,* an instrument that records variations in breathing patterns. Your instructor will advise you how to set up the instrument so you can perform various exercises and read the results. The student should be actively involved in these exercises and should not be giving attention to recordings until completion. It is best to work with partners. One partner can record the physiological notations identifying various exercises for later observation. The pneumograph tubing should be attached firmly but not tightly around the thoracic cavity. Space for chest expansion must be available.

1. Breathe normally for approximately 2 minutes in a sitting position. Inspiration should be recorded by a downward deflection. The pneumograph tubing has been lengthened, thus increasing the volume but decreasing the pressure.

2. Vary your activities in the following manner, and have your partner record the results:

   a. Talking
   b. Standing
   c. Talking and standing
   d. Coughing
   e. Contracting biceps
   f. Deep breathing
   g. Shallow breathing
   h. Run in step
   i. Concentrating
   j. Sitting and extending legs forward
   k. Drinking water
   l. Holding your breath

3. Breathe normally for 2 minutes. Record. Hold your breath as long as you can. Record. Record the recovery period. CO_2 accumulation causes recovery. Recovery usually results when alveolar air reaches 7%–10% CO_2. How long was your recovery period? _____ Average recovery is slightly more than 1 minute. An individual usually breathes 14–18 inspirations and expirations per minute. Now hyperventilate for 45 seconds and then hold your breath. Can you hold your breath longer or shorter now? _____ Explain.

Respiratory System

DISCUSSION

1. a. My tidal volume is _____

    b. My inspiratory reserve volume is _____

    c. My expiratory reserve volume is _____

    d. My approximate residual volume is _____

    e. My total lung capacity is approximately _____

2. What are the pleura and where are they located?

3. How do the fetal pig and human lungs differ with respect to number and names of lobes?

4. Why do the lungs of the fetal pig feel firm rather than spongy?

5. Which cartilage forms the "Adam's apple"? _____

6. Why is it advantageous that the epiglottis be cartilaginous?

7. What is the advantage of the alveolar sacs being simple squamous epithelium?

8. Describe two activities that you believe would produce respiratory variations with the pneumograph.

9. Label, interpret, and attach *your* pneumograph records here.

UNIT X *Nervous System*

A. *Nervous System Anatomy*

PURPOSE The purpose of Unit X-A is to familiarize the student with the histological and gross anatomical features of the nervous system.

OBJECTIVES In order to complete these exercises, the student must be able to do the following:

1. Using a microscope, identify the characteristics of a neuron.
2. Using a microscope, identify the features of the spinal cord.
3. Using a microscope, differentiate between cerebrum and cerebellum tissue.
4. Identify the meninges.
5. Identify the major gross anatomical characteristics of the brain.
6. Identify selected cranial and spinal nerves.

MATERIALS prepared slides of motor neurons (nerve smear)

slides of cross sections of spinal cord, and cerebellum

preserved sheep or beef brains with meninges attached (if available)

ox spinal cord (preserved)

PROCEDURE Neurons, the basic structural and functional units of the nervous system, are characterized in a number of ways: (1) by the direction of impulse flow; (2) by the number of processes; and (3) by the absence or presence of myelinated sheaths.

EXERCISE 1 *Identification of Neuron Types and Cytological Characteristics of Neurons*

Examine a prepared slide of a *nerve smear*. This section probably was prepared from the spinal cord. Carefully focus the slide; then, using highpower or oil immersion, locate large stellate *cell bodies* exhibiting two or more extensions or *processes*. These cells are **neurons**. Neurons with only two processes are *bipolar*, in which one process—the *axon*—carries impulses away from the cell body, and the other—the **dendrite**—carries impulses toward the cell body. *Multipolar* neurons have many processes including one axon and several dendrites. Select a distinct neuron and identify the basic structures (see Figure 109) described below.

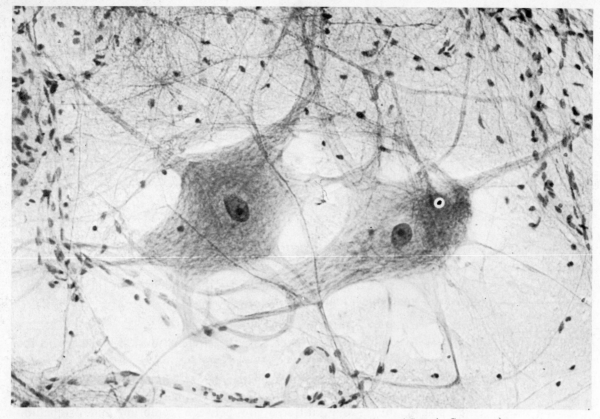

FIGURE 109 *Motor nerve cells of spinal cord* (Courtesy Carolina Biological Supply Company)

The *perikaryon,* or cell body, from which the processes extend, contains a large nucleus, or *karyon,* which contains a rather large nucleolus in addition to other nuclear materials. The cytoplasm or neuroplasm contains *Nissl bodies* and *neurofibrils* in addition to the usual cytoplasmic inclusions. Nissl bodies are granules of RNA and are darkly stained. Neurofibrils are tubular structures; their function has not been identified.

Observe an axon extending from a cell body. Some axons are surrounded by a segmented protective sheath—the *myelin sheath.* Peripheral to the myelin sheath is an additional layer, the *neurilemma,* which plays a role in regeneration of axons. The constrictions between the segments of myelin are termed *nodes of Ranvier,* which are surrounded by the neurilemma. Sometimes a smaller axon branches off the main axon and is termed an *axon collateral.*

EXERCISE 2   *Microscopic Examination of Spinal Cord*

Obtain a prepared slide of a cross section of an ox or human spinal cord. Observe the slide under a dissecting microscope. Notice the central gray matter, which is H-shaped, and the peripheral white matter tracts. Identify the *ventral* and *dorsal medial sulci.* These grooves are on the ventral and dorsal surfaces in the center of the spinal cord. The *dorsal gray columns* or *dorsal horns* are the dorsal extensions of gray matter of the "H," whereas the *ventral gray columns* or *ventral horns* are the ventral extensions of the "H." Fibers of **afferent** *sensory nerves* enter at the dorsal gray columns, and the fibers of **efferent** *motor nerves* extend from the ventral gray columns. The *dorsal median septum* is the smooth white area on either side of the dorsal median sulcus. The *funiculi* are divisions of white matter surrounding the gray dorsally, laterally, and ventrally.

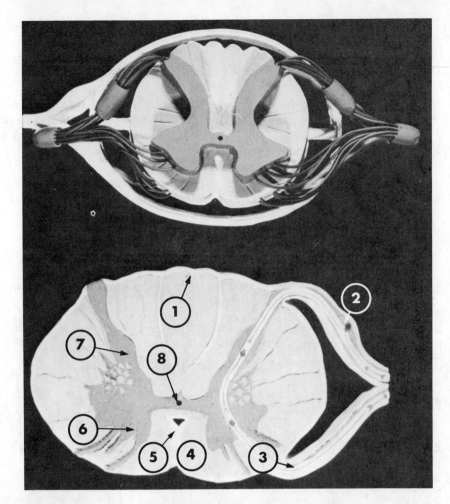

FIGURE 110 *Spinal cord, showing fiber tracts*

1. Dorsal Median Sulcus

2. Dorsal Root Ganglion (Sensory)

3. Ventral Root (Motor)

4. White Matter

5. Ventral Median Fissure

6. Ventral Horn

7. Dorsal Horn (Gray Column)

8. Central Canal

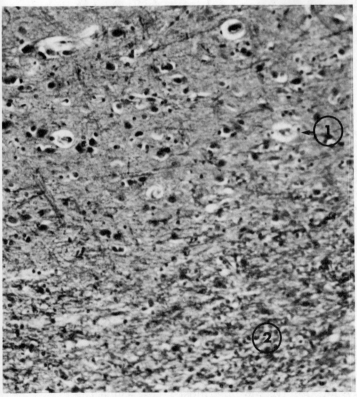

FIGURE 111 *Human cerebral cortex*

1. Pyramidal Cells of Cortex

2. White Matter

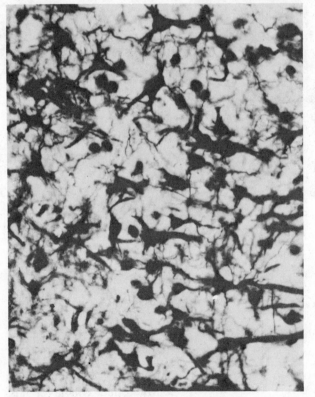

FIGURE 112 *Pyramidal cells as seen in the cerebral cortex*

187

EXERCISE 3   *Microscopic Examination of Cross Sections of Cerebrum and Cerebellum*

In the cerebrum, notice the gray matter of the *cortex* and the white matter of the *corpus callosum.* The obvious *pyramidal cells* constitute most of the gray matter. Long threadlike fibers in the central motor region of the cerebrum originate from pyramidal cells (see Figures 111 and 113).

Examine a cross section of cerebellum. Note the gray matter toward the periphery and the white matter of the **arbor vitae.** Also notice the large **Purkinje cells** at the interface of the cerebellar cortex and medulla (see Figures 112 and 114).

FIGURE 113 *Detail of pyramidal cells in cerebrum*   **FIGURE 114** *Purkinje cells on cerebellum*

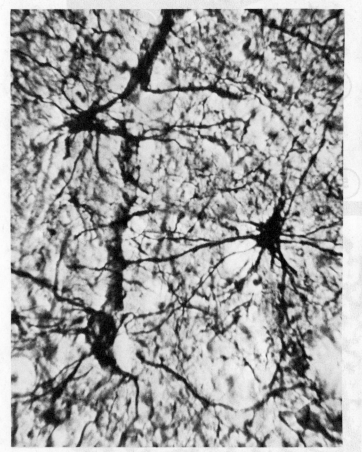

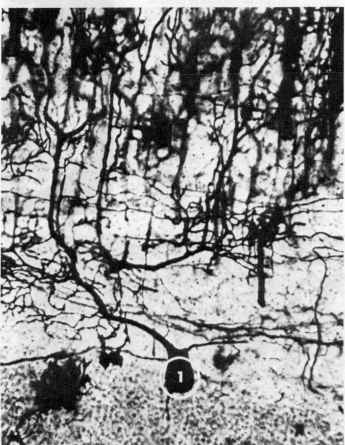

1. Purkinje Cell

EXERCISE 4   *Examination of Meninges*

The **meninges** are protective coverings around the brain and spinal cord. Examine a preserved brain. The outer thick whitish layer is the *dura mater.* Gently make a horizontal incision and then a vertical incision through this layer and reflect it back. Beneath the dura mater, you will see a whitish-to-transparent covering having a spider-web pattern. This is the *arachnoid layer.* Gently reflect this layer laterally and observe the innermost layer, which is adherent to the sulcus and fissures of brain itself. This third innermost layer is the *pia mater.* Observe the spinal cord specimen and locate the meninges if they are present. Between the meninges are spaces called the *subarachnoid* and the *subdural spaces.* The subarachnoid contains *cerebrospinal fluid.* Blood sinuses of the skull are in the dura mater and the dried blood may be obvious on your specimen. Fine vascularization may be obvious in the *pia mater.*

EXERCISE 5   *Examination of Preserved Brain*

The brain of the fetal pig is soft and immature, therefore it is not suitable for dissection. Using a preserved sheep or beef brain, identify the **cerebrum, cerebellum, medulla oblongata, longitudinal cerebral fissure,** and **paired cerebral hemispheres** on the *dorsal* surface. You may be able to identify the **olfactory bulbs,** which extend anteriorly if they remain intact. Ventrally identify anteriorly to posteriorly, the *olfactory bulbs,* and *tracts,* the *optic nerves,* the **optic chiasma,** the **infundibulum,** the **hypophysis** (pituitary gland) and stalk, **mammillary body, pons, pyramidal tract, hemispheres** of the cerebellum, medulla oblongata, and spinal cord.

FIGURE 115 *Sheep brain, superior aspect*

1. Postcentral Gyrus of Cerebrum

2. Cerebellar Hemisphere

3. _____

4. _____

5. Sulci

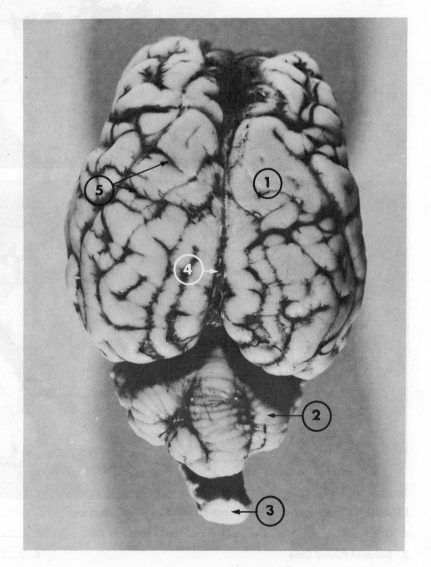

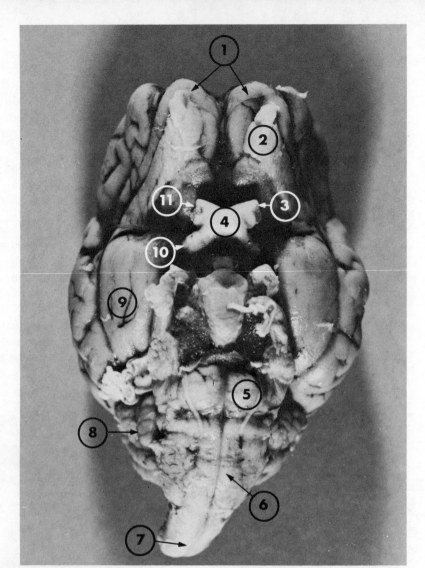

FIGURE 116 *Sheep brain, inferior aspect, showing gross detail*

1. _____

2. Olfactory Bulb

3. Left Optic Nerve

4. Optic Chiasma

5. _____

6. Medulla Oblongata

7. _____

8. Cerebellum

9. _____

10. Optic Tract

11. Right Optic Nerve

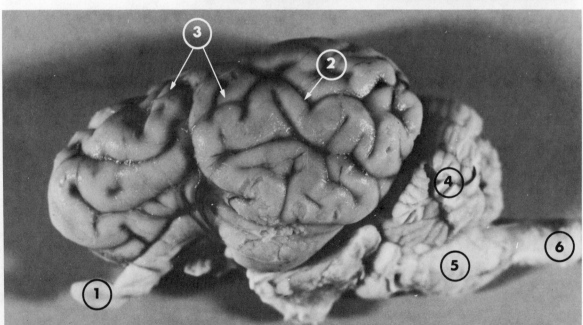

FIGURE 117 *Sheep brain, lateral aspect, showing gross structures*

1. Olfactory Bulb

2. Sulcus of Parietal Lobe

3. Precentral and Postcentral Gyri of Cerebral Hemisphere

4. _____

5. _____

6. _____

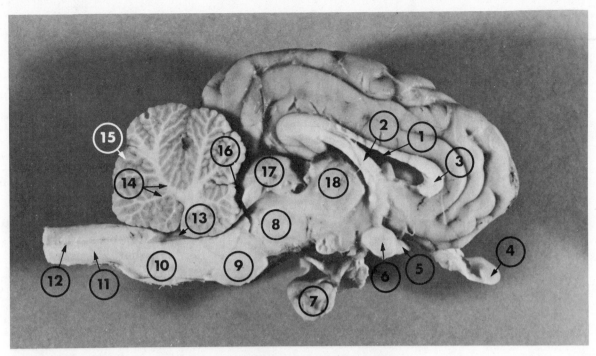

FIGURE 118 *Sheep brain, midsagittal section, showing internal structures*

1. Lateral Ventricle
2. Fornix
3. Corpus Callosum
4. Olfactory Bulb
5. Optic Nerve
6. _Optic Chiasma_

7. _Pituitary_
8. Midbrain
9. Pons
10. _medulla oblongata_
11. _spinal cord_
12. Central Canal of Spinal Cord

13. Fourth Ventricle
14. Medullary Body of Cerebellum (Arbor Vitae)
15. _Gray matter_
16. Third Ventricle
17. Corpora Quadrigemina
18. Thalamus

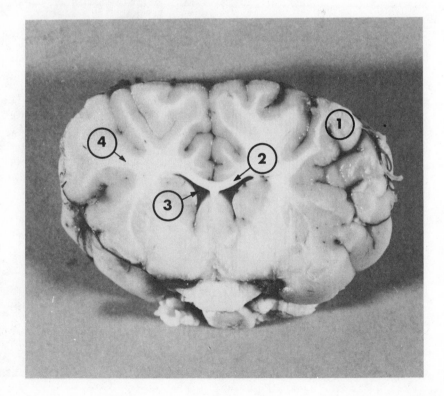

FIGURE 119 *Sheep brain, coronal aspect, showing gray and white matter detail*

1. Gray Matter of Cerebral Cortex
2. Corpus Callosum
3. Lateral Ventricle
4. Medullary Body of Cerebrum (White Matter)

On the *external lateral view* of the brain, locate the *frontal, parietal, temporal,* and *occipital* lobes of the cerebrum. Sulci are deep depressions in the cerebral cortex. Locate the *central sulcus* (fissure of Rolando), which runs superiorly to inferiorly in the central region. Anterior to the central sulcus and running in the same direction is the *precentral gyrus* (which is less deep), and posterior to the central sulcus is the *postcentral gyrus.* Immediately posterior to the frontal lobe on the lateral surface of the brain is the *lateral sulcus.* Make a midsagittal incision through the longitudinal fissure of the cerebrum and continue the incision posteriorly through the cerebellum and spinal cord. Observe the following structures in *median view:* the *corpus collosum, fornix, thalamus, hypothalamus, septum pellucidum, pituitary stalk,* and *pituitary gland.* The *third* and *fourth ventricles* lie between the cerebrum and cerebellum and between the cerebellum and spinal cord, respectively. In the cerebellum locate the *arbor vitae,* and in the spinal cord note the *reticular formation.* Observe the pons, medulla, and spinal cord toward the posterior inferior portion of the brain. In the central region, observe the *intermediate mass, pineal body, mammillary body,* and **midbrain.** Various *cranial nerves* may project partially from the **brainstem.**

FIGURE 120 *Human brain, superior aspect, showing hemisphere detail*

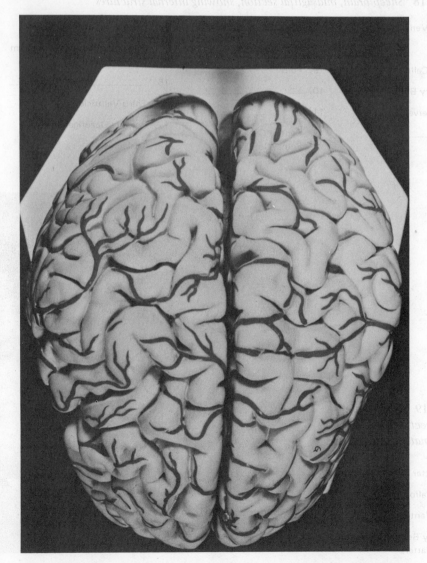

EXERCISE 6   *Examination of Cranial Nerves*

On the preserved brain with intact cranial nerves locate the 12 pairs of cranial nerves, which are:

On   I.   Olfactory
Old   II.   Optic
Olympus III. Oculomotor
Towering IV. Trochlear
Top   V.   Trigeminal
A   VI.   Abducens
Fin   VII. Facial
and   VIII. Acoustic (Vestibulocochlear)
German   IX.   Glossopharyngeal
Viewed   X.   Vagus
A   XI.   Spinal accessory (Accessory)
Hop   XII. Hypoglossal

FIGURE 121 *An isolated complete cat nervous system showing all major nerves, dorsal aspect* (From M. J. Timmons, *A Visual Guide to Dissection: Cat Anatomy Slides*, courtesy J. B. Lippincott Company)

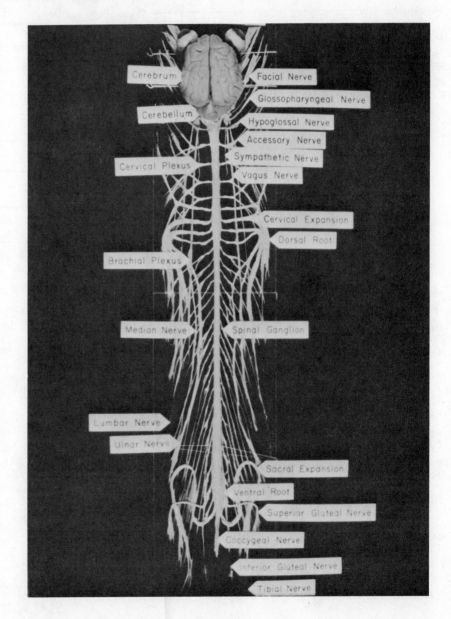

FIGURE 122 *Human brain, midsagittal region*

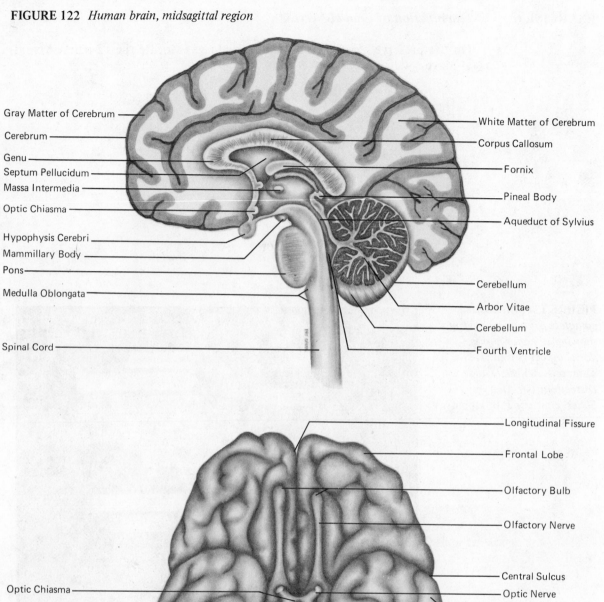

Gray Matter of Cerebrum

Cerebrum

Genu

Septum Pellucidum

Massa Intermedia

Optic Chiasma

Hypophysis Cerebri

Mammillary Body

Pons

Medulla Oblongata

Spinal Cord

White Matter of Cerebrum

Corpus Callosum

Fornix

Pineal Body

Aqueduct of Sylvius

Cerebellum

Arbor Vitae

Cerebellum

Fourth Ventricle

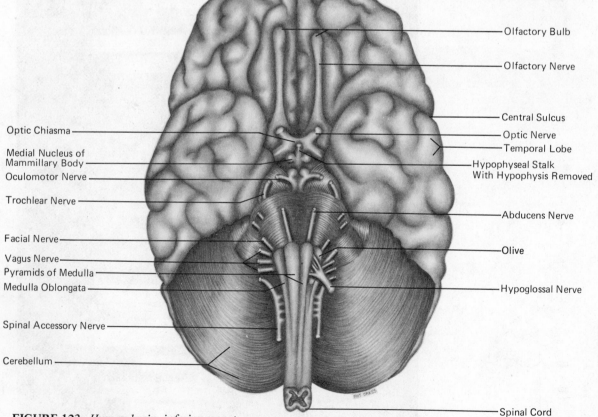

Longitudinal Fissure

Frontal Lobe

Olfactory Bulb

Olfactory Nerve

Central Sulcus

Optic Nerve

Temporal Lobe

Hypophyseal Stalk
With Hypophysis Removed

Abducens Nerve

Olive

Hypoglossal Nerve

Spinal Cord

Optic Chiasma

Medial Nucleus of
Mammillary Body

Oculomotor Nerve

Trochlear Nerve

Facial Nerve

Vagus Nerve

Pyramids of Medulla

Medulla Oblongata

Spinal Accessory Nerve

Cerebellum

FIGURE 123 *Human brain, inferior aspect*

EXERCISE 7   *Examination of Spinal Nerves*

On the preserved ox spinal cord, or demonstration preparation, note the emerging nerves (see Figures 124–126). In the human being, there are 31 pairs: 8 cervical, 12 thoracic, 5 lumbar, 5 sacral, and 1 coccygeal. There are 33 pairs of spinal nerves in the pig: 8 cervical, 14 thoracic, 7 lumbar, and 4 sacral. To expose the **brachial plexus** in your pig, carefully remove the muscles, blood vessels, and connective tissue from the shoulder axillary region, and chest on one side of the body. The brachial plexus, which consists of interconnected white, tough nerves emerging from the last three cervical and first lumbar vertebrae should now be visible. By removing the muscles, blood vessels, and connective tissue from the lumbosacral region, the **lumbosacral plexus** can be seen. It is comprised of branches from the last three lumbar and first sacral nerves.

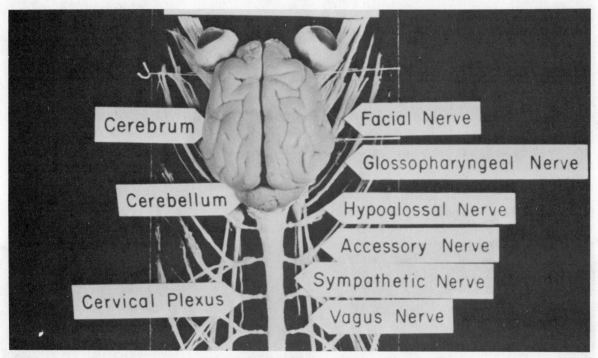

FIGURE 124 *Cat nervous system at the level of the frontal lobe to the cervical plexus showing cranial and spinal nerves, dorsal aspect* (From M. J. Timmons, *A Visual Guide to Dissection: Cat Anatomy Slides,* courtesy J. B. Lippincott Company)

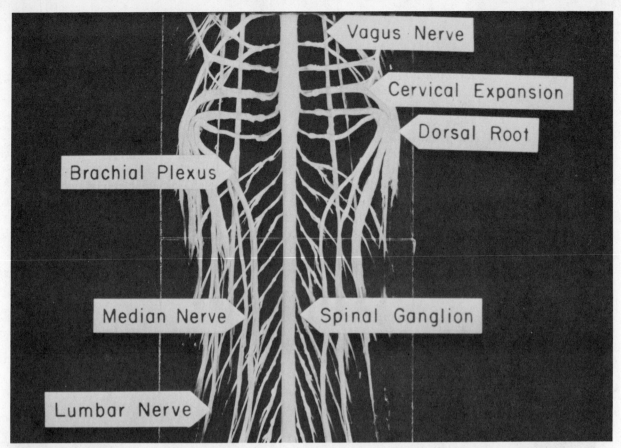

FIGURE 125 *Cat nervous system at the level of the cervical plexus and posterior to the dorsal root showing nerve detail, dorsal aspect* (From M. J. Timmons, *A Visual Guide to Dissection: Cat Anatomy Slides,* courtesy J.B. Lippincott Company)

FIGURE 126 *Cat nervous system posterior to the dorsal root showing nerve detail, dorsal aspect* (From M. J. Timmons, *A Visual Guide to Dissection: Cat Anatomy Slides,* courtesy J. B. Lippincott Company)

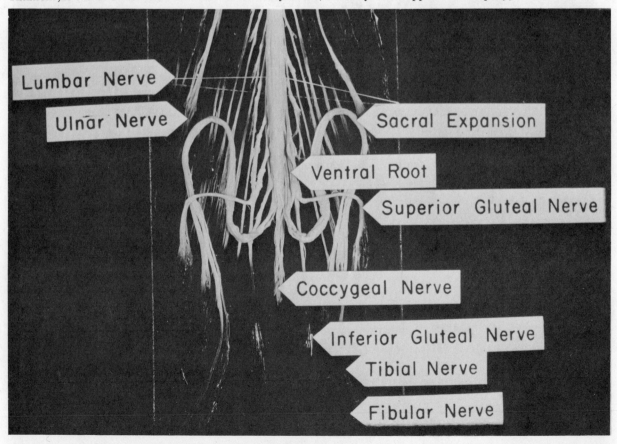

B. *Nervous System Physiology*

PURPOSE   The purpose of Unit X-B is to familiarize the student with some of the physiological aspects of the nervous system.

OBJECTIVES   To complete Unit X-B the student must be able to do the following:

1. Identify selected characteristics of nerve impulse transmission in the frog.
2. Identify selected spinal **reflexes** in the frog.
3. Identify and demonstrate some selected reflexes in the human being.

MATERIALS

live frogs
source of heat
1% HCl solution
Ringer's solution
rubber reflex hammer
pans of water
electric stimulator
kymograph or physiological
  recording apparatus
10% NaCl solution
10% sugar solution
tongue depressors

dissecting instruments
facial tissues
5% acetic acid
flashlight
penlight (optional)
coffee, tobacco, or spices
cotton
cotton-tipped applicator
stopwatch
tuning fork (512 cycles per second)
swivel chair

PROCEDURE   Nerves conduct impulses upon stimulation. For a nerve to be excitable and changed from a resting state, a difference in **ionic** concentration must exist between the external and internal surfaces of the cell membrane. Conductivity results from a continuous change in ionic concentration from the original point of stimulation along the nerve.

EXERCISE 1   *Cranial and Spinal Nervous Responses of the Frog*

In order to demonstrate nerve physiology in the frog successfully, the frog must be freed from pain by *pithing*. Pithing, in contrast to **anesthetizing** the frog, eliminates side effects and produces more valid responses. The frog will be alive, the muscles will contract, and respiration will continue. Two methods of pithing can be utilized, each evoking different responses. *Single pithing* involves destruction of the brain only, while the spinal cord remains intact and functional. In *double pithing,* both the brain and spinal cord are destroyed (refer to Unit V-B, "The Physiology of Muscle Contraction," for pithing instructions).

Perform the following exercises and complete the chart. Work with an unpithed frog first; then repeat with the single-pithed frog and the double-pithed frog. Record your results in the table provided.

1. Pinch the toes of the hind leg. Repeat in 5 minutes.
2. Gently touch cornea (corneal reflex).
3. Apply 5% acetic acid to the Gastrocnemius muscle of one leg.
4. Introduce a light source (flashlight) to eyes and then darken area over eyes with hand (pupillary reflex).
5. Place frog in a pan of water and see if it swims.
6. Place frog on its dorsal surface to see if it turns over to ventral surface.

Exercises	Frog Responses			Conclusions
	Unpithed	*Single Pithed*	*Double Pithed*	
1. Pinching toes	1.	1.	1.	1.
2. Corneal reflex	2.	2.	2.	2.
3. Acetic acid	3.	3.	3.	3.

	Frog Responses			
Exercises	*Unpithed*	*Single Pithed*	*Double Pithed*	*Conclusions*
4. Pupillary reflex	4.	4.	4.	4.
5. Swimming	5.	5.	5.	5.
6. Righting self	6.	6.	6.	6.

EXERCISE 2   *Selected Characteristics of Nerve Impulse Transmission*

A nerve-muscle preparation such as the one you used in Unit VI-B will facilitate your understanding of various characteristics of nerve physiology. In order to make a nerve–muscle preparation, follow the procedure given in Exercise 2 of Unit V-B. Set up the physiological apparatus or **kymograph** according to your instructor's direction and do the following exercises.

a. *Effects of the Application of Various Stimuli*

To the end of a cut sciatic nerve, apply each of the following stimuli, cutting off the destroyed nerve ending after each application (record your data on the chart "Variations in Stimuli"):

1. 1% HCl
2. 0.5% HCl
3. hot water
4. cold water
5. electric shock
6. 1% NaCl solution
7. NaCl crystals
8. pinching of nerve

b. *Conductivity*

Stimulate a freshly cut nerve ending with a mild electric shock and observe the responding muscle contraction. Now remove the nerve-muscle preparation used in part "a. *Application of Various Stimuli*" of this exercise and place it in a petri dish, keeping it moist with Ringer's solution. Soak a small amount of cotton in Novocain. Remove the nerve from the Petri dish and apply cotton to it at a point approximately 1 cm from its cut end. Allow cotton to remain on the nerve for 60 seconds. Set up the nerve-muscle preparation as before. Stimulate the nerve with a mild electric shock in a region anterior to the anesthetized area. Record results. Repeat stimulation at the point where Novocain was applied. Record your results. Repeat stimulation at a point posterior to Novocain application.

VARIATIONS IN STIMULI

Stimulus	Type of Stimulus: chemical, osmotic, electrical, mechanical	Response
1. 0.5% HCl	1.	1.
2. 1% HCl	2.	2.
3. Hot water	3.	3.
4. Cold water	4.	4.
5. 1% NaCl	5.	5.
6. NaCl crystals	6.	6.
7. Mild electric shock	7.	7.
8. Strong electric shock	8.	8.
9. Pinched nerve	9.	9.

EXERCISE 3　　　*Reflexes in the Human Being*

　　　　a. *The Swallowing Reflex*
Swallow the saliva in your mouth and immediately swallow again. Explain your result and compare it to the rapid succession of swallowing demonstrated by rapidly drinking a glass of water.

Try to stop yourself from swallowing. Explain in terms of a reflex action.

　　　　b. *The Patellar Reflex*
Sit so that your leg from the knee hangs down freely. Have another student strike the patellar ligament (just below the knee) with a rubber hammer. This may require a little patience, and it is best for the subject to divert his attention. Notice that the leg is extended by the contraction of the quadriceps muscle group. Repeat on another student. Is the reflex obtained just as readily and is it equally extensive in all students? Record your data.

　　　　c. *Photo–Pupil Reflex*
Close your eyes for 2 minutes; while facing a bright light, open them and let another student examine the pupils immediately. Describe the observed response. What is the purpose of this reflex action?

d. *The Accommodation Pupil Reflex*

In a moderate light, look at a distant object (20 ft or more removed) and have another student examine your pupils. Now look at a pencil held about 10 in. from the face (without changing the illumination) and note the pupils. What is the purpose of this reflex?

e. *Convergence Reflex*

Look at a distant object. Have another student note the position of the eyeballs. Look at a near object. What change is observed in the eyeballs? This is convergence. What is the purpose of this reflex?

f. *The Achilles or Ankle Jerk*

Kneel on a chair; let the feet hang freely over the edge of the chair. Bend your foot so as to increase the tension of the Gastrocnemius muscle. Have your partner tap with a rubber hammer the tendon of Achilles. What reflex results?

g. *Corneal Reflex*

Touch the cornea of your eye with a piece of facial tissue. What is the result? What is the purpose of this reflex?

EXERCISE 4   *Tests for Cranial Nerve Function in the Human Being*

A superficial assessment of cranial nerve function can be made by performing relatively simple procedures. Working in pairs, test each other's cranial nerve function in the following manner. Record your results below as you perform each test and then fill in the table in Question 12 of the Discussion at the end of the unit.

a. *Olfactory Nerve (Cranial Nerve I)*
Ask the subject to identify the odor of coffee, tobacco, or spices. The test may not be valid if the subject has a cold. Additional tests of olfactory discrimination will be performed in Unit XI, Exercise 3.
Result:

b. *Optic Nerve (Cranial Nerve II)*
Ask the subject to read a portion of a printed page with each eye, wearing glasses if necessary. Additional vision tests will be performed in Unit XI, Exercise 1.
Result:

c. *Oculomotor Nerve (Cranial Nerve III)*
Ask the subject to follow your finger, a pencil or a penlight with his eyes, keeping his head still as you slowly move it up, then down.
Result:

d. *Trochlear Nerve (IV) and Abducens Nerve (VI)*
Have the subject follow your finger, a pencil, or a penlight with his eyes, keeping his head still, as you slowly move it laterally in each direction. As well as innervating eye movements through the extrinsic rectus ocular muscles, cranial nerves III, IV, and VI also innervate the upper eye lid and provide parasympathetic stimulation to the pupils. Therefore, you should observe your subject for signs of **ptosis** (drooping of one or both eyelids), and, with the room lights darkened, for reaction to light, if you have not done so in Exercise 3c. To test for reaction to light, better results will be obtained if you bring the penlight (or flashlight) in from the side, rather than the front of the subject. Observe the pupil. *Result:*

e. *Trigeminal Nerve (Cranial Nerve V)*

To test the motor responses of this nerve, ask the subject to clench his teeth. With the examiner providing resistance by holding his hand under the subject's chin, ask the subject to open his mouth. He should be able to do both.
Result:

To test the sensory responses of this nerve, have the subject close his eyes. Test for light touch by whisking a piece of dry cotton over the mandibular, maxillary, and ophthalmic areas of the face. Wet the cotton with cold water, and determine whether the subject is able to discriminate temperature in these same areas.

With the subject's eyes open, gently touch a piece of dry cotton to the cornea, if you have not done so in Exercise 3g.
Result:

f. *Facial Nerve (Cranial Nerve VII)*

To test the motor response of this nerve, ask the subject to wrinkle his forehead, raise his eyebrows, puff his cheeks, and smile showing his teeth. Look for any asymmetry.
Result:

To test the sensory response of this nerve, touch a cotton-tipped applicator stick which has been dipped into a 10% NaCl solution to the tip of the tongue. Repeat with 10% sugar solution on the anterior surface of the tongue. Additional exercises involving taste will be done in Unit XI, Exercise 4.
Result:

g. *Acoustic Nerve (Cranial Nerve VIII)*

To test the cochlear portion of this nerve, determine the subject's ability to hear a ticking stop watch and repeat a whispered sentence. The Weber and Rinne tests (see Unit XI, Exercise 2), which utilize a tuning fork, may also be done at this time.
Result:

To test the vestibular portion of this nerve have the subject sit in a swivel chair or stool, turn him ten turns in approximately twenty seconds, then stop the chair suddenly. Observe the subject's eyeballs for rapid movement or quivering, known as **nystagmus**, which is a normal finding when one is dizzy. Nystagmus is not normal under nonexperimental conditions and may indicate a disorder of the inner ear or of cranial nerves III, IV, or VI.

In general, if the subject is able to keep his balance while walking, the vestibular branch of cranial nerve VIII is all right.
Result:

h. *Glossopharyngeal Nerve (IX) and Vagus Nerve (X)*
These nerves may be tested together. If you wish (and your subject is willing), test the gag reflex by touching the subject's uvula with a cotton-tipped applicator.
Result:

The motor portion of these nerves may be tested by: (1) asking the subject to swallow some water; (2) holding his tongue down with a tongue depressor and asking him to say "ah" (the uvula should move); and (3) noticing any hoarseness when he speaks.
Result:

i. *Spinal Accessory Nerve (Cranial Nerve XI)*
This procedure tests the strength of the Trapezius and Sternocleidomastoid muscles, which are innervated by this nerve. To check the strength of the Trapezius, place your hands on the subject's shoulders and determine whether he can raise them against resistance. To test the strength of the Sternocleido-mastoid muscle, place your hands on each side of the subject's head and ask him to turn his head to each side against resistance.
Result:

j. *Hypoglossal Nerve (Cranial Nerve XII)*

Ask the subject to "stick out" his tongue. The tongue should protrude straight, with no deviation.

Result:

EXERCISE 5   *Tests for Cerebellar Function in the Human Being*

The cerebellum functions to maintain coordination, posture, and gait. After asking the subject to perform each of the following, check off whether he was able to perform the task (+) or was not able to perform it (−).

Test	+	−
a. With eyes closed, touches index finger of each hand to nose		
b. With eyes open, touches examiner's fingers		
c. Moves hands and fingers fast		
d. Looking straight ahead, moves the heel of one foot down the shin of the other leg		
e. Looking straight ahead, touches his outstretched hand with corresponding toe		
f. Stands with feet together and eyes closed without losing balance		
g. While walking, arms swing slightly		
h. While looking straight ahead, walks in tandem (heel to toe) without losing balance		

Nervous System

DISCUSSION

1. Compare the cytological and physiological characteristics of an axon and dendrite.

2. Suggest reasons why mature neurons are not replaced after destruction:

3. Where do afferent (sensory) nerve impulses enter the spinal cord? Where do efferent (motor) nerve impulses leave the spinal cord?

4. Give the function of the following:

   a. olfactory bulbs _____

   _____

   b. pyramidal tracts _____

   _____

c. optic chiasma _____

_____

d. thalamus _____

_____

e. hypothalamus _____

_____

f. cerebellum _____

g. arbor vitae _____

5. What is a spinal reflex? Give an example:

6. Where do the pupillary and corneal reflexes have their centers of control?

7. Describe a subliminal, liminal, and maximal stimulation of a nerve-muscle preparation.

8. How does a local anesthetic influence nerve impulse conductivity?

9. Which ions are involved in impulse transmission?

10. What is the purpose of the various reflexes you observed?

11. Attach recorded results and briefly explain (from Exercise 2).

12. Complete the following chart for cranial nerves.

Number	Name and Branches	Sensory or Motor	Function

UNIT XI *Special Senses*

PURPOSE

PURPOSE   The purpose of Unit XI is to familiarize the student with the basic anatomy of some of the sense organs and with the physiology of the eye, ear, nose, taste buds, and cutaneous receptors.

OBJECTIVES In order to complete Unit XI, the student must be able to do the following:

1. Identify structural components of the eye.
2. Perform physiological exercises related to vision.
3. Identify anatomical features of the ear.
4. Perform physiological exercises related to hearing.
5. Perform physiological exercises related to olfactory discrimination.
6. Perform physiological tests to indicate various taste discriminations.
7. Identify histological characteristics of some cutaneous receptors.
8. Perform exercises illustrating physiological characteristics of cutaneous receptors.

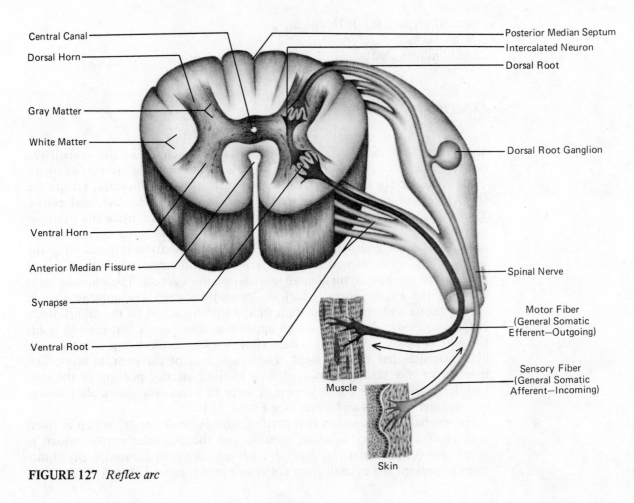

FIGURE 127 *Reflex arc*

One commonly thinks of five senses: vision, hearing, taste, smell, and touch. The senses also include other sensations—those of warmth, cold, and pain.

The sense organs, which include all the sensory neurons, are involved in experiencing or reacting to any of these senses or sensations. These organs serve two functions—that of sensation and that of reflex—both of which result from the stimulation of sense **receptors**. Sensory receptors are classified into three main groups:

1. Exteroceptors: Surface receptors (dendrites) in the skin, eye, and ear.
2. Visceroceptors: Receptors in blood vessels, stomach, and intestines.
3. Proprioceptors: Receptors in muscles, tendons, joints, and the internal ear (to detect body positioning and movement).

The nerve endings primarily contained in the skin are referred to as **cutaneous** receptors.

EXERCISE 1 *The Eye*

MATERIALS

preserved sheep or cow eyes
Snellen charts
pins
source of bright light or flashlight
tape measures
color blindness charts

PROCEDURE

DISSECTION OF THE EYEBALL

Make a coronal section through the preserved specimen of the eyeball and gently place it on your dissecting tray. It is important that you keep the contents relatively intact to facilitate identification of structures. Locate the three layers or coats covering the eyeball, the **sclera**, **choroid**, and **retina**. These three layers are beneath the *extrinsic* muscles. Identify the extrinsic muscles: the *superior rectus, inferior rectus, medial rectus, lateral rectus, superior oblique,* and *inferior oblique.* The sclera is often referred to as the white of the eye. Note that the anterior portion or **cornea** is transparent and covers the **iris** or pigmented portion of the eyeball. The choroid layer contains the *ciliary body,* which is located between the anterior margin of the retina and posterior margin of the iris. Attached to the ciliary body is the *suspensory ligament,* which appears as transparent threads and holds the *lens* in place. The lens is a hard (but flexible in the living state) kernel-like structure that refracts light. The *iris* is part of the choroid layer. The inner layer—the retina—is located only in the posterior portion of the eyeball. It appears to be a thin brownish layer of tissue containing photoreceptor neurons—the *rods* and *cones* (see Figure 131).

The eyeball also contains two cavities: the *anterior cavity,* which is filled with clear and watery *aqueous humor:* and the *posterior cavity,* which is larger and contains *vitreous humor,* a gelatinous viscid substance that functions to prevent the eyeball from collapsing (see Figures 128 and 129).

FIGURE 128 *Sheep eye, midsagittal section, showing gross internal detail*

1. Sclera

2. Retina

3. _____

4. _____

5. _____

6. Conjunctiva

7. Vitreous Humor

8. _____

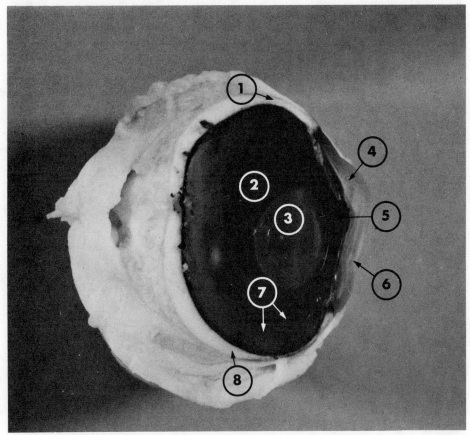

FIGURE 129 *Sheep eye, coronal section, showing internal detail, posterior to lens*

1. _____

2. Fovea (Macula Lutea)

3. Retina

4. _____

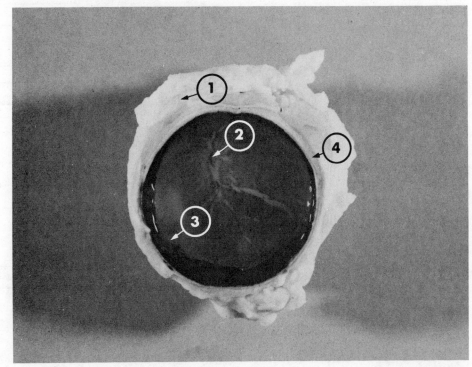

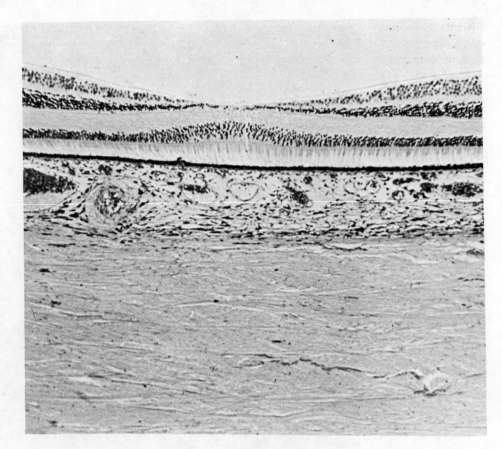

FIGURE 130 *Fovea centralis area of retina*

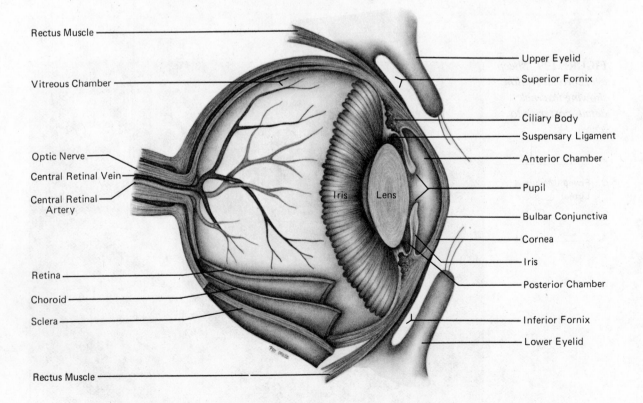

FIGURE 131 *Human eye, midsagittal section*

VISION

The eyes function in the image-formation aspect of vision and in the stimulation of nerve impulses, which are conducted to visual areas of the cerebral cortex. The image upon the retina is formed by refraction of light rays as they move through the eyeball from the cornea to the retina. Rods and cones are dense toward the center of the retina where vision is most acute. This region is called the *fovea*. Toward the periphery, the photoreceptors become less dense and visual acuity decreases. At the point where the *optic nerve* enters the retina, there is a complete absence of rods and cones. This region is referred to as the *blind spot*. Examine prepared microscope slides of the eye (see Figure 130).

The following exercises exemplify some characteristics of the eye and vision:

1. Following your instructor's directions, read the letters on the Snellen chart. Be sure to cover one eye. You can determine the approximate strength of vision in both your right and left eyes. Record. Normal vision is 20/20, which means that an individual being tested can read letters at a distance of 20 feet that a person with normal vision can see at 20 feet. If vision were 20/50, the person being tested could see at 20 feet what an individual with normal vision could see at 50 feet. The measurements 20/50, 20/20, and so on, are specific to either the right or the left eye.

2. You can determine the approximate distance from your eye that an image falls onto the blind spot of each eye by doing the following exercises.

    a. To determine the blind spot in your left eye, hold this page 18 inches away from you and, closing your right eye, focus on the circle. Slowly bring the page closer, still focusing on the circle. The point at which the cross disappears is the blind spot where the optic nerve enters your right eye. Measure this distance and record it in the following table.

    b. To determine the blind spot in your right eye, hold this page 18 inches away from you and, closing your left eye, focus on the cross. Slowly bring the page closer, still focusing on the cross. The point at which the circle disappears is the blind spot where the optic nerve enters your right eye. Measure this distance and record it in the following table.

TABULATION OF CLASS NORMS FOR COMPARISON

	Distance from Eye at Which Blind Spot Becomes Evident:	
Person Tested	*Right Eye*	*Left Eye*

Class norm:

3. Examine the various color blindness charts available. Be sure to follow the instructor's directions in reference to each chart.

EXERCISE 2   *The Ear*

MATERIALS

stop watch   tuning forks: (1) 500 hertz; (2) 2000 hertz
rubber reflex hammer   (cycles per second)

PROCEDURE

The auditory apparatus is composed of the ears, auditory nerves, and auditory areas of the temporal lobes of the cerebrum. Each ear consists of (1) an external ear; (2) a middle ear; and (3) an internal ear. The external ear has two divisions: (1) the *auricle* or *pinna*—the flap of tissue on the outside of the ear; and (2) the *external* or auditory *acoustic meatus*—a canal terminating at the *tympanic membrane* or *eardrum*. The tympanic membrane is a tense membrane of circular and radial fibers which separates the external ear from the middle ear. This membrane transforms sound waves into mechanical vibrations, which are transmitted into the middle and inner ear. The tympanic membrane also serves as a barrier that protects the delicate auditory ossicles of the middle ear.

The *middle ear* or *tympanic cavity* contains three tiny bones called *auditory ossicles.* They are the *malleus* or hammer; the *incus* or anvil; and the *stapes* or stirrup. These bones transfer and magnify sound waves to the fluid within the cochlea of the inner ear.

The *inner ear* is concerned with both hearing and the sense of equilibrium. The hearing organ itself is the *organ of Corti* and is contained within the **cochlea** portion of the inner ear (see Figures 132 and 133). Hearing results

FIGURE 132 *Section of cochlea*

1. Helicotrema
2. Scala Media
3. Organ of Corti
4. Scala Tympani
5. Scala Vestibuli
6. Spiral Ganglion
7. Cochlear Nerve

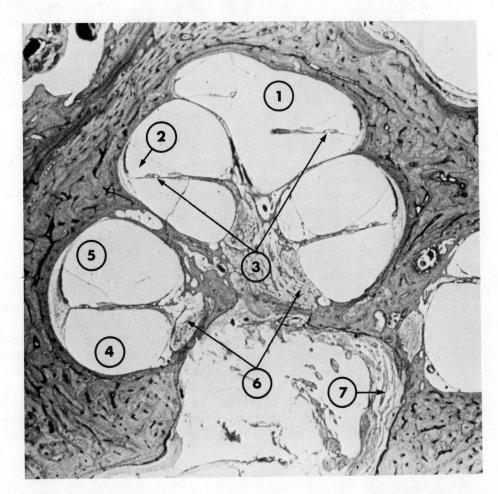

FIGURE 133   *Organ of corti*

from stimulation of the auditory area of the cerebral cortex, located in the temporal lobe of the brain. Equilibrium or balance is sensed by the flow of fluid within the three *semicircular canals* in the inner ear.

The ear transmits pattern of sound vibrations, their intensities, and directions of origin to the temporal lobe. Sound waves have two major characteristics: *frequency,* or wavelength, which determines pitch; and *intensity,* or height of waves, which determines loudness. The following exercises are designed to illustrate characteristics of hearing.

A. WATCH-TICK TEST

Work with a partner. Hold a stop watch very close to your partner's ear. Slowly move away from his or her ear until the tick is no longer audible. Repeat with the other ear. Chart your results in inches after measuring when the sound becomes inaudible. Tabulate class data for comparison.

B. WEBER AND RINNE TESTS

A tuning fork can be stimulated to vibrate with a rubber reflex hammer and is used to compare hearing in both ears. It is also possible to differentiate between conductive and perceptive (or sensineural) deafness using a tuning fork. To do the *Weber* test, strike the tuning fork with the hammer and place the handle on your forehead in a medial position. If the tone is heard in the middle of the head, you have equal hearing or loss of hearing in both ears. If nerve deafness is present in one ear, the tone will be heard in the other ear. In conduction deafness, the sound will be heard in the ear in which there is hearing loss. In the *Rinne* test, strike the tuning fork and place the handle on the mastoid process. Then place the tines parallel to the external auditory meatus. Alternate the fork in the two positions a number of times. If the tone was heard equally well in both places, hearing loss is probably mixed. If hearing if normal, the tone should be louder and heard longer at the side of the ear (air conduction). If it is louder in the back (bone conduction), a conductive hearing loss may be indicated.

C. EXAMINATION OF PREPARED SLIDES

Examine prepared microscope slides that show the middle and inner ear structures.

EXERCISE 3   *Olfactory Discrimination*

MATERIALS

Samples of 10 spices, such as nutmeg, thyme, cinnamon, sage, vanilla extract, allspice, mustard powder, potato, ginger, rosemary, garlic powder, pepper, oregano, clove, paprika, and slices of apple and potato.

PROCEDURE

Receptors for the sense of smell are situated high in the interior of the nose between the median **septum** and in the region of the superior turbinate. This area is referred to as the **olfactory** *cleft*. Olfactory nerves pass through the cribriform plate (above the olfactory cleft) of the ethmoid, which is lateral and inferior to the crista galli and from there pass to the nasal interior (see Figure 103).

The nasal passages are lined with mucous membranes and cilia, which condition the air in the breathing process. When one has a cold, the membranes tend to swell and secrete excessive mucous, which covers the lining of the cavity and, therefore, impairs the sense of smell.

The following are experiments designed to test your ability to discriminate different smells. The sense of smell is complex and abilities to discriminate various smells are individual. Test your ability to discriminate and compare with the class.

1. Class groups of 10 students are suggested for this exercise. Your instructor has selected several different spices. You are to smell each with your eyes closed and identify the spice. Keep a list of the numbered spice sample and your response.

2. Pinch your nostrils together, close your eyes and have your partner place a piece of either potato or apple in your mouth. Is it apple or potato? Explain.

EXERCISE 4   *Taste*

MATERIALS

prepared slides of tongue—
   cross sections
menthol eucalyptus
cotton-tipped applicators
PTC paper

sugar, lemon, burnt almonds, salt,
   pepper, mustard
ice cubes
hot H_2O

PROCEDURE

Taste, like smell, is result of sensory neuron stimulation by chemicals. The tongue, which is covered with papillae (see Figures 68 and 69), is the principal organ of taste. Taste buds (see Figure 67) are spread over the surface of the tongue and are sensitive to substances dissolved in water. The four tastes are *salty,* stimulated by metallic cations; *sour,* stimulated by hydrogen ions; *sweet,* stimulated by a hydroxyl group in sugar or alcohol; and *bitter,* triggered by alkaloids.

1. Your instructor will have various solutions in jars, which you can apply to your tongue systematically. Dip the applicator in one solution and gently apply the applicator to the tongue in the following positions: (1) apex; (2) posterior medial region; (3) anterior lateral sections; and (4) posterior lateral sections. Map on the drawing below where you can most readily identify the solution. Repeat the applications with other solutions, mapping your results. Compare your mapped responses with other class members.

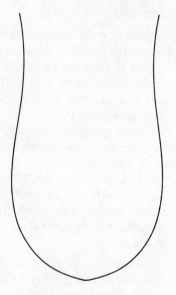

2. Next, hold an ice cube on your tongue for a minute. Apply a sugar solution to your tongue. What is the result?

3. Next, hold hot water in your mouth so the tongue is exposed to the hot water. Remove the water and apply your sugar solution to the tongue. What is the result?

4. Place a piece of PTC paper in your mouth. The ability to taste the chemical in the paper is inherited. Compare your ability to taste the chemical with your classmates.

   What percentage of the class are "tasters"?_____

   What percentage of the class are "non-tasters"? _____

5. Examine prepared microscope slides of cross sections of the tongue showing taste buds.

EXERCISE 5   *Cutaneous Receptors*

MATERIALS

prepared slides of skin and sense receptors	brushes with firm bristles
	pins
powdered charcoal	ice
hot water	centimeter rulers

PROCEDURE

The following exercises will familiarize you with a variation of sensations related to the skin.

A. TOUCH

Review the cross section of the skin, noting the layers of the epidermis and their histological characteristics. Note the depth and inclusions of the dermis as well. Your instructor has available a selection of sensory neurons found in the skin, such as *Pacinian* and *Meissner's corpuscles*. What are the various sensations stimulated by these neurons?

1. Mark off an area with a pen 2 cm square in size on the palm of the hand and on the inner surface of the wrist. Use 0.5-cm markings to make a grid in this 2-cm area on the palm and wrist.
2. With your eyes closed, have your partner explore the marked off areas with the bristle or needle (lightly) using the same intensity of stimuli over 20 different spots within the 2-cm areas of the palm and wrist. If a sensation is felt, mark in the grids a "T" for touch.
3. Have your partner place a small piece of cork on the skin of the forearm of the same arm that was used in the above steps, and determine how long it takes for the initial sensation of pressure to give way to an indifferent sensation. Repeat and record the time required to reach this accommodation to sensory stimuli.

B. TEMPERATURE RECEPTORS

With your eyes closed, have your partner map the cold and hot receptors in the grid area of your palm and inner wrist. Use a dissecting needle or finishing nail that has been cooled in ice water, wipe dry and again explore in the same manner 25 different spots within the grids. Mark "C" for every cold receptor located within the grid. After mapping the wrist and palm area for cold receptors, wait 5 minutes before exploring for hot receptors. Repeat the previous procedure, this time using a dry hot finishing nail or your dissecting needle (be careful not to burn yourself!). Again record the hot receptors with "H" in the grid areas of the palm and wrist areas.

Repeat procedures A and B, using the calf of the leg or bottom of the foot. Compare the differences in receptors of the palm, wrist, calf, and/or foot according to function.

C. DISCRIMINATION OF STIMULI

Close your eyes and have your partner apply two pins or needles or bristle points to the palm of your hand and determine how far apart the two points must be before you can discriminate two distinct points of stimuli. Your partner should test your response by touching, now and then, only a single point to the skin. Repeat the above procedure to the following areas: index finger (palmar surface), forearm (with grid), back of the neck, and calf or feet (plantar surface). Record your results for the above areas of stimulation in terms of millimeters of separation.

D. DISTRIBUTION OF SWEAT GLANDS

In order to demonstrate the distribution of these glands, the palm and wrist of the subject's hand must be towel dried. Your partner will now lightly dust your forearm and the palm of your hand with powdered charcoal. *Use the hand that has the grid marked on the palmar surface.* After dusting with the charcoal powder, wait 15 seconds and, with a sharp puff, blow the excess powder from the hand and wrist. Where sweat glands are present, the powder will adhere to the moisture produced by the glands. Count the dots in the 2-cm-square areas of the palm and wrist. Record. Are there any differences in the numbers of glands present? If time permits, you may wish to do other areas of the body in this same manner, such as the bottom of the foot (plantar surface).

Special Senses

DISCUSSION

1. What structures of the eyeball contribute to refraction?

2. What does 20/20 vision mean?

3. What is the difference between a conductive and a perceptive hearing loss?

4. What structures are associated with maintenance of equilibrium?

5. Why are the Weber and Rinne tests diagnostic?

6. Is the ability to taste approximately the same for all individuals? *Explain.*

7. Where are the main sensory neurons associated with smelling located?

8. Why is smelling impaired during upper respiratory infections?

9. What conditions can influence variations in ability to sense touch?

10. What types of color blindness are most common?

UNIT XII *Urinary System*

PURPOSE
: The purpose of Unit XII is to familiarize the student with the general anatomy and physiology of the urinary system and to enable the student to understand the basic tests included in a routine urinalysis.

OBJECTIVES
: In order to complete Unit XII, the student must be able to do the following:

1. Identify anatomical structures of the urinary system in a male and female fetal pig.
2. Identify gross anatomical features of the kidney.
3. Using a microscope, identify the glomerulus and nephron unit.
4. Demonstrate accuracy in performance of basic tests included in a routine urinalysis.

MATERIALS

sheep or beef kidneys	Clinistix	No. 2 filter paper
male and female	Clinitest	Petri plates
fetal pigs	Albustix	nitric acid (conc.)
microscope slides of	Ictotest	microscope
renal tissue	Ketostix	
urinometers	Hemastix	
blank microscope slides	Combistix	
cover slips	hot plate	
nitrazine paper	10% acetic acid	
unknown urine samples	Benedict's reagent	
test tubes		
400-ml beakers		

PROCEDURE

EXERCISE 1
: *Examination of Urinary Organs of the Male and Female Fetal Pig*

This exercise may be completed on your fetal pig specimen, but be sure to observe a specimen of the opposite sex as well. The urinary organs are located in the pelvic region. The kidneys are covered and attached to the dorsal body wall by a tough serous membrane termed the *peritoneum*. Remove or pull aside the small intestine, colon, and pancreas in order to provide an unobstructed view of the urinary apparatus. Carefully remove the fat, connective tissue, and peritoneum that cover the ventral aspect of the kidneys and blood vessels (renal arteries and veins, descending aorta and postcava) that supply this organ.

The field should be sufficiently exposed to make a detailed examination of the urinary apparatus. The urinary system is made up of a pair of **kidneys**, a pair of **ureters**, a **urinary bladder**, and **urethra** (see Figure 135). Examine and identify the following surfaces of each kidney: superior, inferior, medial, lateral, dorsal and ventral. Identify the creamy orange **adrenal glands** which lie on the superior and medial borders of each kidney. These are endocrine glands which secrete various hormones that influence metabolism, cardio-vascular output, and water-salt balance. The hormones secreted by these glands have physiological influences on the kidneys but are not a functional component of the urinary system. Identify the medial border of each kidney and note the depressed or indented portion termed the **renal hilum**. This is the point where the renal blood vessels, nerves, and ureter enter and exit the kidney. Identify the **ureters** which exit from the hilum and lie against the dorsal body wall. Trace the ureters inferiorly from the kidneys to the point of attachment with an elongated, collapsed sac which lies between the umbilical arteries. This sac is the **allantoic bladder** or fetal **urinary bladder**. Dorsally the bladder narrows as embryonic development proceeds to form a short canal termed the **urethra** which will carry urine from the urinary bladder to the outside environment.

Carefully lift up the right kidney and, with your scalpel, slit the kidney with a longitudinal cut from the lateral border to the medial border. Remove the cut half of the kidney and observe the position with the vessels still attached. What are the vessels? Identify the hilar region where the ureter exits from the kidney. Trace the **renal artery** (may have several branches) from the abdominal aorta to the renal hilum. Follow the **renal vein** from the postcaval vein to the renal hilum. In Exercise 2 you will observe the gross internal features of a kidney.

FIGURE 134 *Sheep kidney, midsagittal section, showing internal detail*

1. Medullary Region

2. Calyx

3. Renal Pelvis

4. _____

5. Renal Pyramids

6. Renal Column

7. Cortex

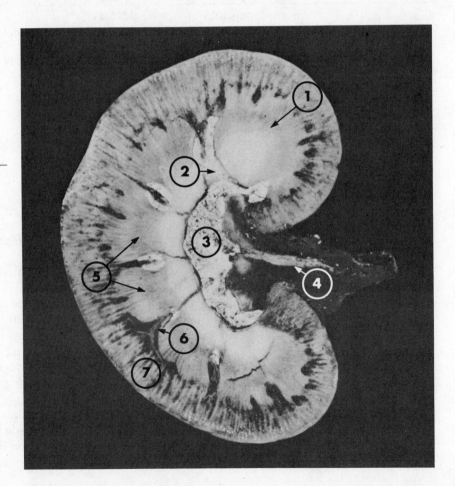

EXERCISE 2   *Observation of the Gross Anatomy of a Kidney*

Use a preserved beef or sheep kidney for this exercise. Carefully remove the excessive adipose tissue. Be careful not to cut the ureter or any attached renal arteries or veins. The adipose tissue holds the kidney in position in a living animal and is very important. Observe the fibrous connective tissue covering that surrounds and protects the kidney. It is called the *renal capsule*. Also note the **hilum**, which is the indentation where the ureter and

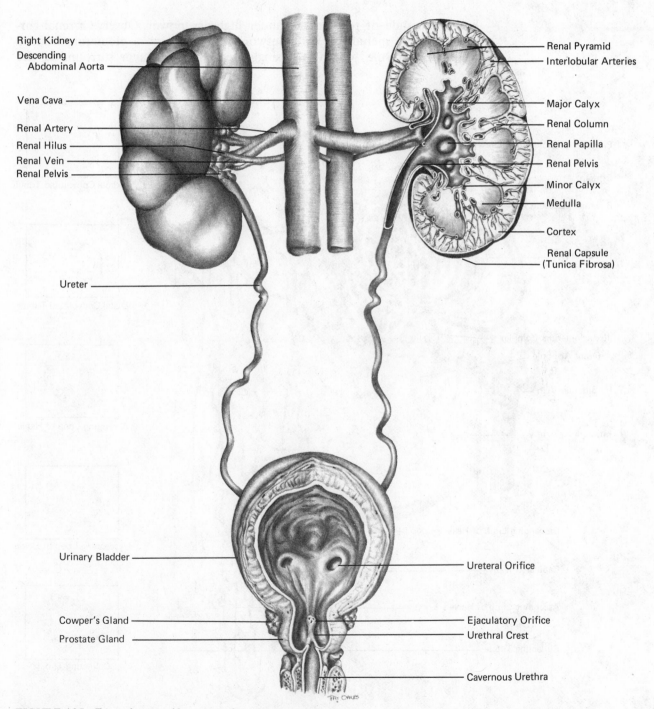

FIGURE 135 *Frontal view of human male urinary system*

blood vessels enter. Make a coronal section through the kidney starting at the outer convex region. Open the kidney halves and observe the following structures: **renal cortex**; **renal medulla**, including **pyramids, medullary rays, papillae, calyces,** and **columns**; and the **renal pelvis** (see Figure 134 for reference). Note that the renal pelvis is formed by a wall of thick white fibrous tissue and forms the orifice of the ureter.

EXERCISE 3   *Microscopic Examination of Renal Tissue*

Focus the slide of renal tissue under high dry power. Observe a **renal corpuscle**, which includes a tight network of capillaries—a **glomerulus**—and a **Bowman's capsule**, which is a single epithelial cell layer surrounding the

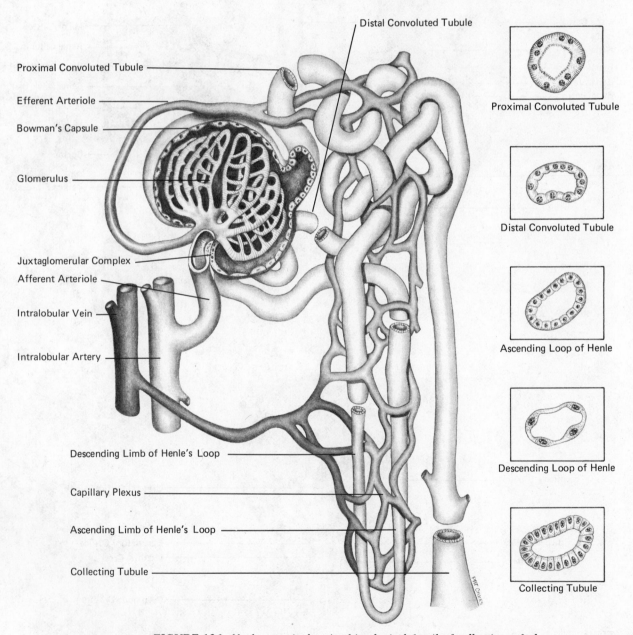

FIGURE 136 *Nephron unit showing histological detail of collecting tubules*

glomerulus. Follow a Bowman's capsule as it narrows and extends from the renal corpuscle. This twisted tubule is the **proximal convoluted tubule**. Note that the tubule continues in an inferior direction and then turns superiorly forming a loop. This tubule section is referred to as the **loop of Henle**. As the loop of Henle continues and extends from the renal corpuscle, it is referred to as the **distal convoluted tubule**. As the distal convoluted tubule continues, note that the walls become slightly thicker. These thicker-walled, straight, tubular extensions are called **collecting tubules**. A **nephron**, which is the structural and functional unit of the kidney, is composed of a glomerulus; a Bowman's capsule, and the proximal convoluted, loop of Henle, distal convoluted tubules, and collecting tubules (see Figures 136–139 for reference).

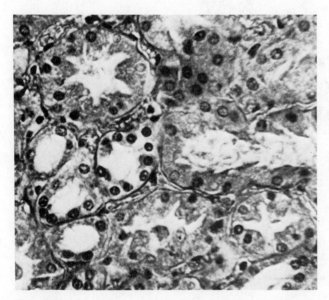

FIGURE 137 *Kidney tubules, cortical region*

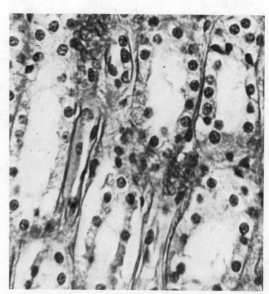

FIGURE 138 *Kidney tubules, medullary region*

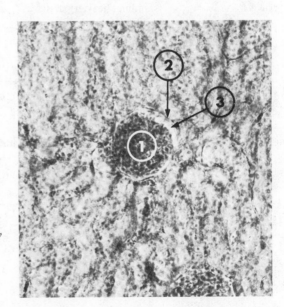

FIGURE 139 *Glomerulus of kidney*

1. Glomerulus
2. Bowman's Capsule
3. Bowman's Space

EXERCISE 4   *Routine Urinalysis*

Urine is approximately 95% water and includes the following dissolved substances: pigments; electrolytes (for example, sodium and potassium); hormones, such as estrogens; and the nitrogenous wastes—urea, uric acid, creatinine, and ammonia. During illness or disease, other substances such as bacterial toxins can be found, as well as abnormal substances such as blood, sugar, and protein. Urine is, therefore, a good general indicator of an individual's state of health. More complete urinalyses are done with a 24-hour urine collection, which is normally about 1500 ml in volume, but routine examinations are done on samples of lesser amounts.

After cleansing the area surrounding the urinary meatus, collect from yourself a small sample—approximately 30–50 ml—of urine. The first urination of the day is best, but it is often difficult to collect in a school situation. Observe the physical characteristics of your urine and record in the table below.

PHYSICAL CHARACTERISTICS OF URINE

Characteristic	Normal	Your Urine	Unknown
Color	Amber to straw colored; can vary with diet, medication, or amount eliminated		
Odor	Fresh—no odor; ammonia odor develops after standing because of bacterial breakdown and formation of ammonia substances		
Turbidity	Clear; becomes cloudy upon standing; heavier particles settle to bottom		
Specific gravity	1.003–1.030 in normal adult; can be higher if an early morning specimen; specific gravity is temperature dependent, so add 0.001 for each 3°C above 20°C; subtract 0.001 for each 3°C below 20°C.		
Microscopic examination	Can include epithelial cells, leukocytes, pigment, casts		

In order to determine specific gravity, use the urinometer or hydrometer and follow the directions specific to the measuring instrument. In order to observe the urine microscopically, make a smear on a clean slide and observe. You may see some epithelial cells sloughed off from the lining of the urinary tract, some white blood cells, some parasites, and/or casts. *Casts* are collections of substances such as inorganic materials or mucus that have hardened and assumed the shape of the tubule in which they were formed.

Also included in routine urinalysis are tests to determine the absence or presence of substances that are abnormal constituents of urine. These tests can be completed accurately and effectively by use of various dipsticks that are on the market. Directions must be followed for each of the specific tests and the results must be compared with the available charts on either the bottle or box. In order to complete the following common tests, dip the required treated strip of paper into your specimen, wait the required time, and record the results in the table below. For the Clinitest and Ictotest, the procedure is different. Follow the directions on the label for these tests. Your instructor will provide you with an "unknown" urine specimen. Analyze both your own and the "unknown" at the same time. Abnormal components may have been added to the unknown.

TESTS FOR ABNORMAL CONSTITUENTS OF URINE

Indicator	Substance Indicated	Results Your Urine (+: presence of substance; -: absence)	Results Unknown
Clinistix	Glucose	_____	_____
Albustix	Albumins, (protein)	_____	_____
Ketostix	Ketones	_____	_____
Hemastix	Blood	_____	_____
Combistix	More than one substance	_____	_____
Ictotest	Bilirubin	_____	_____
Clinitest	Glucose	_____	_____
Nitrazine paper	pH (normal range pH is 4.8–8.0; average is pH 6; may vary with diet and medication)	(give pH) _____	_____

EXERCISE 5   *Urine Screening Tests*

   a. *Test for Urine Protein*
Normal adult urine may contain a trace of plasma proteins (2 to 8 mg/100 ml). Plasma proteins are normally prevented from entering the glomerular filtrate by the glomerulus and by their large molecular weights. An increase in protein of 0.5 to 4 g per day indicates evidence of glomerular disease.

Put 10 ml of urine in a test tube. Then put the test tube into a beaker of tap water and onto a hot plate. Bring the contents of the tube to a boil. As soon as the urine boils, add three drops of 10% acetic acid to the test tube. If a precipitate results, then albumin and globulins are present; this indicates a positive test for proteins.
Results?

List and describe four diseases that may be indicated by *proteinuria.*

1. _____

   _____

2. _____

   _____

3. _____

   _____

4. _____

   _____

b. *Benedict's Qualitative Method for Detecting Urinary Glucose*

Benedict's is a copper reduction test that is used for the qualitative screening of glucose in urine. If glucose is present in urine it reduces Benedict's reagent from blue alkaline copper sulfate to a precipitate of red cuprous oxide. If a yellow, green, orange, or brick red precipitate forms, a reducing sugar (e.g., glucose) is present in the specimen. The production of the various colors depends upon the concentration of reducing sugar present. The brick red color indicates a maximum concentration.

Place 0.5 ml or eight drops of urine into a clean dry test tube. Add 5 ml of Benedict's reagent. Mix by agitating the tube. Place the test tube into a beaker of tap water and onto a hot plate. *Let boil for 5 minutes.* Remove from the boiling water and observe immediately for any color change.

Color Interpretation	Glucose Concentration	Results
(a) Clear blue or green with no precipitate	0 to 100 mg/100 ml	0
(b) Green with a yellow precipitate	100 to 500 mg/100 ml	1+
(c) Yellowish green with a yellow precipitate	500 to 1400 mg/100 ml	2+
(d) Brownish orange with a yellow precipitate	1400 to 2000 mg/100 ml	3+
(e) Orange to brick red precipitate	2000 or more mg/100 ml	4+

Results:

_____

_____

List and describe three diseases that are indicated by *glycosuria.*

1. _____

   _____

2. _____

   _____

3. _____

   _____

c. *Test for Urinary Bilirubin*

Bilirubin results from the breakdown of hemoglobin within the liver. Free bilirubin within the plasma normally does not pass through the glomerulus and is not found in the glomerular filtrate. A positive test for urinary bilirubin may indicate hepatic damage.

Place a small strip of No. 2 filter paper in a Petri dish or watch glass. Add two drops of urine and one drop of concentrate nitric acid (HNO_3). *Please Use Caution.* A green color is positive for biliverdin. Bilirubin is not stable in urine and is oxidized to biliverdin. Other colors are negative. Results?

List and describe three disorders that are indicated by high levels of urinary bilirubin and/or biliverdin.

1. _____

   _____

2. _____

   _____

3. _____

   _____

d. *Test for Urea*

Normal adult urine contains large amounts of urea. The major portion of dietary nitrogen is excreted in the form of urea in the urine.

Place two drops of urine on a clean dry microscope slide. Carefully add two drops of concentrated nitric acid (HNO_3). Slowly warm the slide on a slide warming tray or hot plate. *Do not allow it to boil.* Cool. When crystals begin to form, examine under the low power of your microscope. The crystals that form are urea nitrate. *Draw your results.*

List, describe, and microscopically identify three kinds of crystals which may be found in normal acidic or alkaline urine.

1. _____

   _____

2. _____
   _____

3. _____
   _____

Urinary System

DISCUSSION

1. List the structures through which urine must travel starting with the glomerulus and terminating at the external urethral meatus.

2. Why is it important that Bowman's capsule be one cell layer in thickness?

3. What is the function of adipose tissue surrounding the kidney?

4. Most of the nephron structures are located in what part of the kidney?

5. How does the urinary system of the male differ from the female?

6. Turbidity of urine and specific gravity are related. Why?

7. Why is urinalysis considered a valuable diagnostic test?

8. Which nerves influence urine formation?

9. Which hormones influence urine formation?

10. Identify the function of each component of the nephron.

UNIT XIII *Reproductive System*

PURPOSE   The purpose of Unit XIII is to enable the student to understand the anatomy and physiology of the male and female reproductive systems.

OBJECTIVES   In order to complete Unit XIII, the student must be able to do the following:

1. Identify the structures of the male and female fetal pig reproductive systems.
2. Observe specimens of human or bovine **gonads** (the ovary and testis), penis, placenta, and uterus with fetus in situ.
3. Identify the stages in oogenesis and spermatogenesis.
4. Using a microscope, recognize sections of penis, uterus, ovary, and testis.
5. Understand the basic principles of reproductive physiology.

MATERIALS

fetal pigs	Holtfreter's solution (optional)
dissecting microscope	gross specimens of prepared ovaries, testis,
compound microscope	penis, uterus with fetus in situ, and placenta
pituitary extract (optional)	microscope slides of ovaries, testis, penis,
dissecting instruments	uterus, and sperm smears
live frogs	Petri dishes (optional)

PROCEDURE

EXERCISE 1   *Fetal Pig Reproductive Organs*

a. *Female Reproductive System*

Observe the reproductive organs of the female fetal pig (See Figure 140). Determine the pubic symphysis and with your scalpel cut through it so that the hindlimbs can be reflected laterally. It may be necessary to cut away part of the pubis, which is still cartilaginous. This will allow for better exposure. Again exercise care *not* to cut deeper than necessary.

Above the tail and anus, identify the **urogenital orifice**. This structure is external. Immediately above the orifice and still external, note the slight elevation or **genital papilla**.

Now referring to Figure 140, dissect the **urethra** away from the **vagina**, gently cutting through the connective tissue. Make a midline incision through the vagina and extend the incision anteriorly through the uterus. Lay open the vagina and uterus. Then extend the incision posteriorly to the **clitoris**, a slight elevation above the genital papilla.

The anterior region of the uterus is the **body** which tapers slightly posteriorly to form the **cervix**. The vagina is continuous with the uterus and like

239

the uterus is a muscular tube. The posterior section below the vagina is the **urogenital sinus,** the reservoir common to both urinary and reproductive systems.

From the midline and the body of the uterus, the uterine horns branch laterally. Note the **broad ligament** which attaches the uterine horns to the abdominal wall dorsally. Also note the **abdominal ostium** a continuation of the **Fallopian tubes,** or uterine tubes extending anteriorly above the small, yellow-white, bean-shaped **ovaries.** The **abdominal ostium** serves as the point of entry for the ova produced by the ovary. Identify the **mesovarium,** or mesentary linking the ovaries to the posterior aspect of the broad ligament.

If available, compare the fetal uterus to a pregnant pig uterus. Note the **placenta** in the pregnant uterus, which allows for the exchange of materials between mother and fetus. In the pig, note the **chorion** or innermost layer which surrounds the fetus, and the **uterine mucosa** or the uterine lining comprising the placenta.

Carefully cut through the chorionic layer and expose the fetuses. Make the incision in the region of the uterine horns. Identify the **amnion,** the thin sac-like covering of each fetus. Note that the **umbilical cord** of each fetus attaches to the **allantois,** the thin inner layer of the amnion which is heavily vascularized.

b. *Male Reproductive System*

Observe the **scrotum processus vaginalis,** which is the sac below the umbilical cord. The scrotum contains the testes, the male reproductive organs, so it will be necessary to make an incision at the midline into the scrotum. It may be necessary to probe away any muscular or connective tissue that would

FIGURE 140 *Pelvic basin exposed to show female reproductive organs*

1. Oviduct, Left

2. Ovary, Left

3. Uterine Horn

4. Body of Uterus

5. Urinary Bladder, Cut Edge

6. Umbilical Artery

7. Femoral Artery, Left

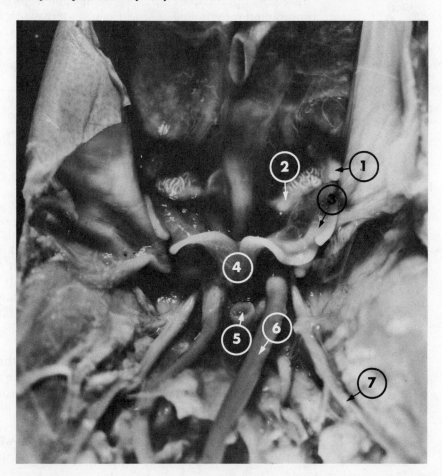

obstruct your observations. If necessary, flush the area because coagulated blood often is found within.

Note the **gubernaculum,** the white connective tissue cord which connects the testes to the scrotum. The collection of tubules surrounding the posterior, dorsal, and anterior surface of the testes is the **epididymis.**

The anterior continuation of the epididymis is the **ductus deferens (vas deferens).** Trace this structure medially. Note that the ductus deferens is encased in a connective tissue covering the **spermatic cord.** Also included in the cord are the spermatic artery, spermatic vein, and nerve.

The two **seminal vesicles,** glands producing semen, can be identified at the point where the right and left ductus deferens enter the urethra. Below the seminal vesicles can be found the small prostate gland which also produces semen. It may be difficult to identify the prostate gland because of the immature condition of the pig. Note the **bulbourethral** or **Cowper's glands** which are paired and elongated. They are located at the posterior junction of the urethra and penis.

The **penis** is a thin, cordlike, muscular structure lying immediately posterior to the urinary bladder. Note that the urethra in the region of the Cowper's glands runs into the penis. It will be necessary to separate the penis from the bladder and umbilical artery. These structures are covered by skin, and joined by connective tissue. They extend to the umbilical cord. Note the external urethral orifice below the umbilical cord.

EXERCISE 2   *Observation of Gross Specimens*

Your instructor will provide you with specimens of **ovary, testis, penis, uterus,** and **placenta** for observation.

EXERCISE 3   *Microscopic Study of an Ovary*

The sexually mature ovary contains millions of undifferentiated cells termed **oogonia,** which, like the spermatogonia, undergo many divisions of mitosis in order to increase the number of these cells. As the oogonia cells mature, these cell types develop into larger **primary oocytes.** The primary oocytes are large cells that contain yolk. These cells divide by **meiosis** into two cell types: the **secondary oocyte,** which contains all yolk, and the **first polar body.** Meiosis continues as the secondary oocyte divides to give rise to a large **ootid** and a smaller **second polar body.** The ootid undergoes further minor changes until it is a mature **ovum** (egg). The polar bodies are formed in order to eliminate the extra number of chromosomes. Thus the ovum contains half the number of chromosomes as the diploid oogonium (see Figure 141). The process by which the mature ovum is formed is called **oogenesis.**

Observe a slide of a sectioned ovary showing **Graafian follicles** (see Figure 142). Notice the *cuboidal epithelium* around the periphery of the section. Medial to the cuboidal epithelium you may see **primary follicles.** After puberty, under hormonal influence, several follicles begin to grow prior to ovulation and become primary oocytes, secondary oocytes, and, later Graafian follicles. Graafian follicles occur only in mammals and are hollow sacs containing an ovum surrounded by **follicular fluid** (or liquor folliculi).

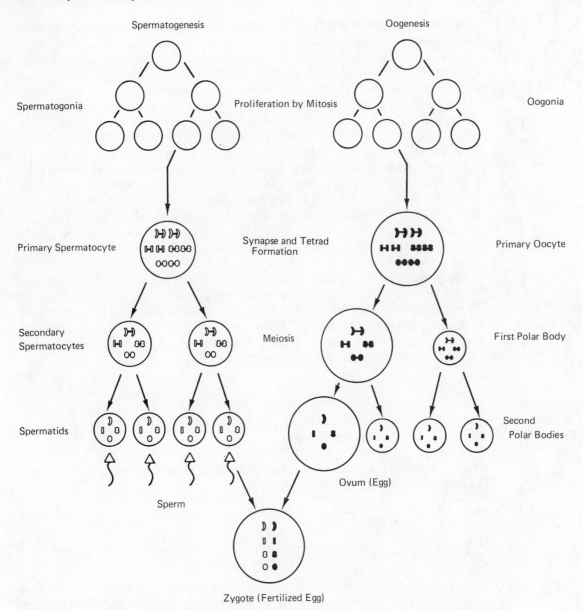

Spermatogenesis

Oogenesis

Spermatogonia

Proliferation by Mitosis

Oogonia

Primary Spermatocyte

Synapse and Tetrad Formation

Primary Oocyte

Secondary Spermatocytes

Meiosis

First Polar Body

Spermatids

Second Polar Bodies

Sperm

Ovum (Egg)

Zygote (Fertilized Egg)

FIGURE 141   *Gametogenesis*

FIGURE 142 *Ova nest with follicle cells as viewed in the cortex of the ovary*

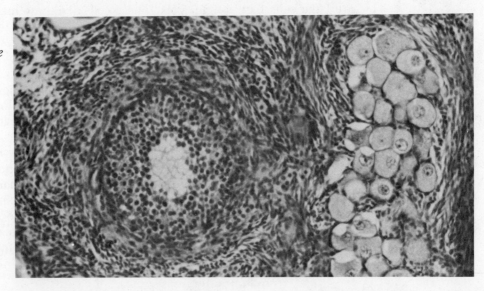

The cavity containing the follicular fluid is the *antrum*. Immediately underlying the ovum within the Graafian follicle is a mound of cells known as the **cumulus oophorus** (see Figure 143). *Make a sketch* of what you see and *label* the various structures.

Observe a slide of a sectioned ovary showing corpora lutea. A **corpus luteum** is a collapsed follicle that has expelled its ovum. Corpora lutea first appear yellow and then regress into scar tissue known as the **corpus albicans**. The active corpus luteum produces *progesterone*, a hormone essential to pregnancy (see Figure 144). Sketch in the corpus luteum and corpus albicans in the above drawing and label these structures.

EXERCISE 4   *Microscopic Study of a Testis*

The testis consists of thousands of coiled tubules that develop millions of sperm. These tubes are lined with undifferentiated germ cells termed **spermatogonia**. These spermatogonia divide by the process of mitosis in order to increase the number of these cells. At the age of sexual maturity, the spermatogonia cells begin to undergo **spermatogenesis**. The first stage of this process is the growth of these cells into much larger cells termed **primary spermatocytes**. These cells in turn divide by **meiosis** into two cells of equal size, termed **secondary spermatocytes**. Meiosis continues as the secondary spermatocytes immediately divide into four equal-sized cell types termed **spermatids**. Spermatids contain half the number of chromosomes as diploid spermatogonia and require further minor changes before becoming functional **spermatozoa** (see Figure 141).

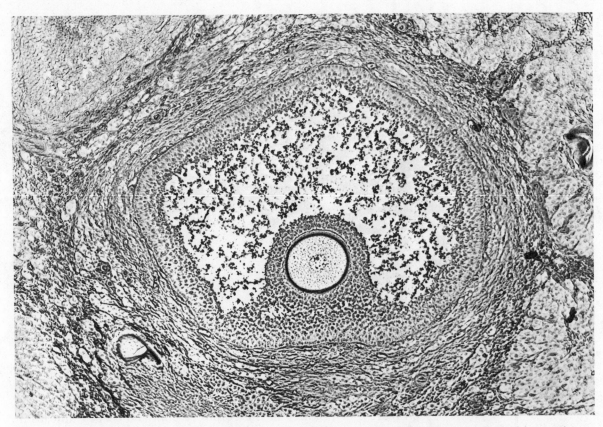

FIGURE 143   *Graafian follicle showing cumulus oophorus* (Courtesy Carolina Biological Supply Company)

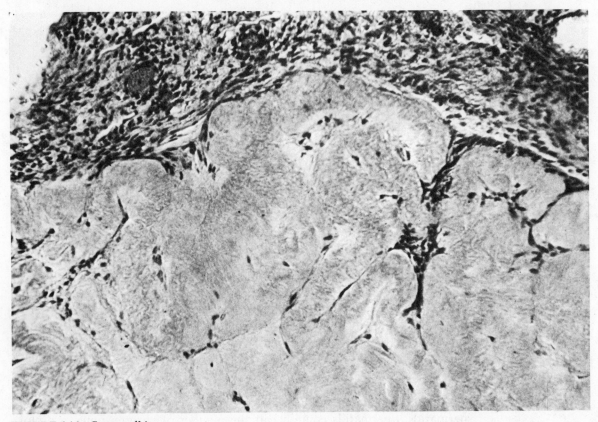

FIGURE 144   *Corpus albicans*

Observe a slide containing a section of a testis. Within the testis, you will find numerous **seminiferous tubules** (see Figure 145). Under high power, focus on a cross section near the periphery of a seminiferous tubule. Starting at the periphery and moving towards the center, you should be able to see the maturation of sperm (spermatogenesis) from spermatogonia to primary spermatocytes to secondary spermatocytes to spermatids to mature **sperm** in the center of the tubules (see Figure 146). Mature sperm can be recognized by the presence of tails. Between the seminiferous tubules are the **interstitial cells of Leydig,** which produce and secrete the male hormone **testosterone.** *Draw and label* a cross section of a testis.

Observe a slide of a sperm smear. *Identify* the head and tail of a spermatozoan.

EXERCISE 5   *Microscopic Study of a Penis*

Obtain a slide of a cross section of a penis. This slide will be best observed under a dissecting microscope. *Identify:* two **corpora cavernosa,** corpus cavernosum urethrae (also called **corpus spongiosum**), and the *urethra,* which is found in the corpus spongiosum. Also observe the connective tissue, muscle, sinuses, and blood vessels in this section.

EXERCISE 6   *Microscopic Study of a Uterus*

Obtain a section of uterus (any phase) and identify the smooth muscle area and endometrial lining. In the **endometrium,** which is sloughed off during **menstruation,** look for arteries and uterine glands. *Sketch and label* this slide.

FIGURE 145 *Seminif-*
erous tubules of testis

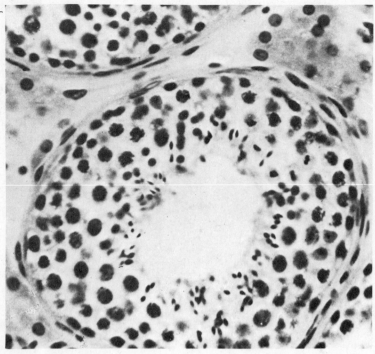

EXERCISE 7   *Induction of Ovulation in the Frog (Optional)*

In this exercise, you will inject frogs with pituitary extract containing **gonadotropic hormones**. Your instructor will provide you with instructions and materials for doing this.

FIGURE 146 *Detail of seminiferous tubule of rat testis showing spermatogenesis*

EXERCISE 8   *Embryology of the Frog (Optional)*

1. First transfer several unfertilized eggs (from Exercise 7) to a Petri dish containing Holtfreter's solution (see Appendix) which should completely cover the eggs. Place a few eggs on a slide, cover with water, and observe the **fertilization membrane** under low power. Examine the remaining eggs under a dissecting microscope and *draw* a few eggs, paying particular attention to their pigmentation. The animal pole of the egg is highly pigmented, whereas the vegetable pole is not.

2. Observe fertilized frog eggs in various stages of embryological development. Identify the following developmental stages of the **embryo**: **cleavage** (2-, 4-, 8-, and 16-cell stages), **blastula**, **gastrula**, and **neurula** (see Figures 147–160). *Draw* the stages of development.

FIGURE 147 *Fertilized frog egg*

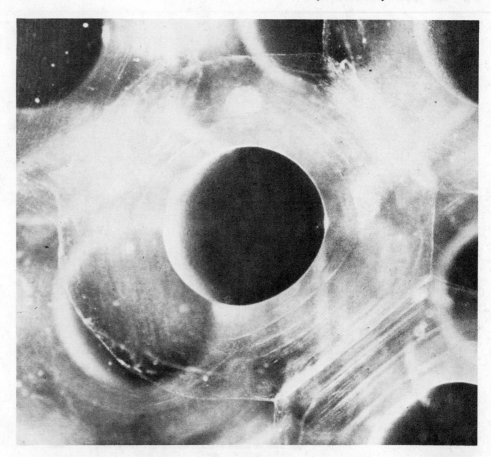

FIGURE 148 *Frog egg, two-cell stage*

Figures 147–160 courtesy Carolina Biological Supply Company

FIGURE 149 *Frog egg, four-cell stage*

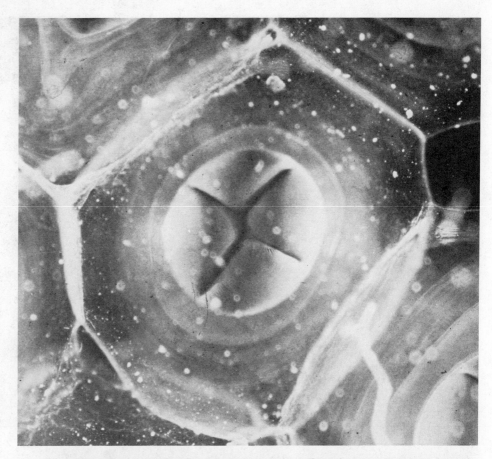

FIGURE 150 *Frog egg, eight-cell stage*

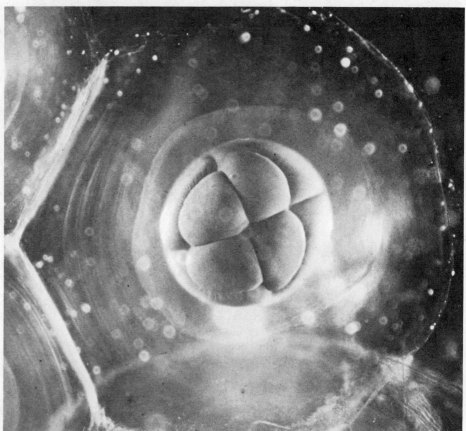

FIGURE 151 *Frog egg, 16-cell stage*

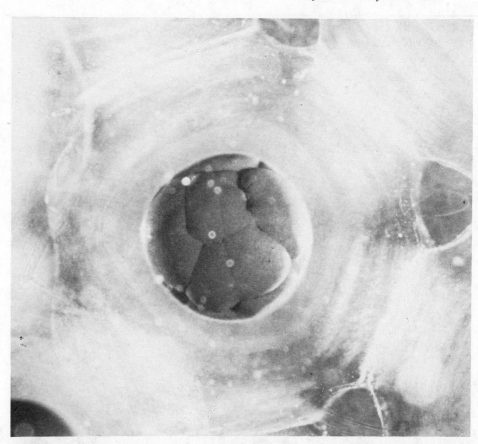

FIGURE 152 *Frog egg, 32-cell stage*

FIGURE 153 *Frog egg, early blastula stage*

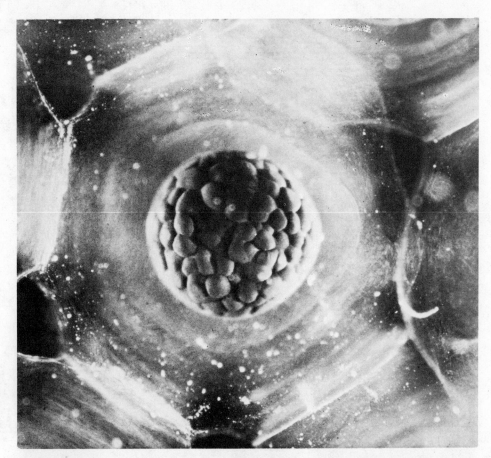

FIGURE 154 *Frog egg, late blastula stage*

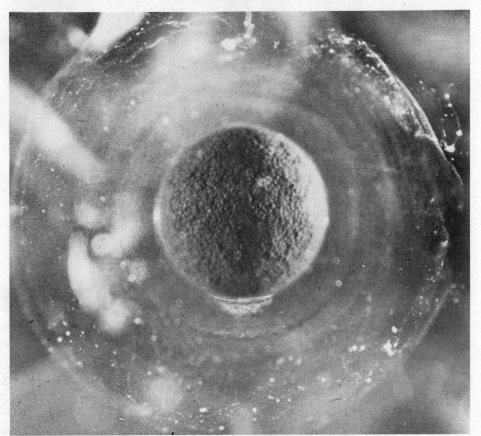

FIGURE 155 *Frog egg, early yolk plug stage*

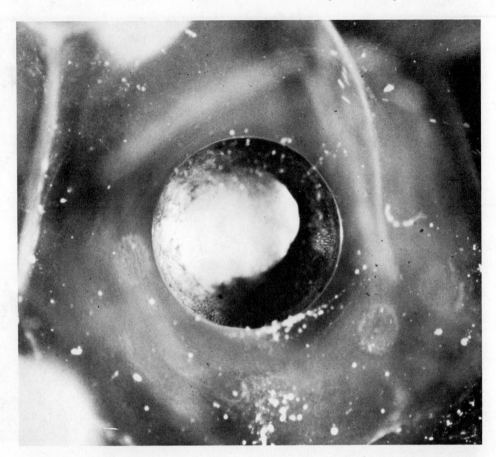

FIGURE 156 *Frog egg, late yolk plug stage*

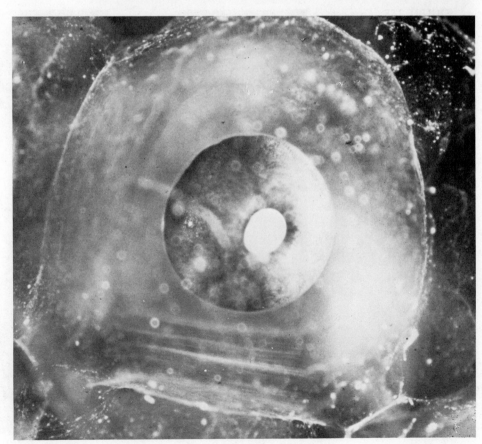

FIGURE 157 *Frog egg,
neural plate stage*

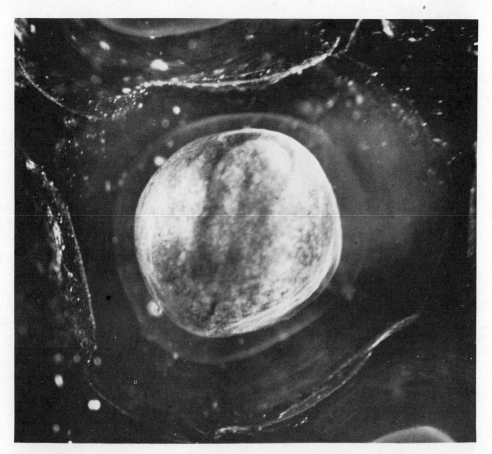

FIGURE 158 *Frog egg,
early neural groove stage*

FIGURE 159 *Frog egg, late neural groove stage*

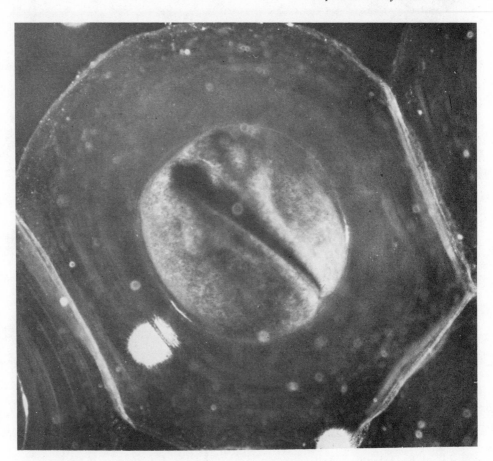

FIGURE 160 *Frog egg, neural tube stage*

Reproductive System

DISCUSSION

1. Trace the passage of sperm from the seminiferous tubules to the point of ejaculation.

2. a. Trace the passage of an egg from its release from a Graafian follicle through implantation.

   b. What is the fate of a Graafian follicle after ovulation?

3. Indicate whether each of the following contains the diploid or haploid number of chromosomes:

   a. sperm _____

   b. primary oocyte _____

   c. secondary spermatocyte _____

   d. Leydig cell _____

   e. ovum _____

4. In the human, which hormone(s) are responsible for inducing ovulation? _____

_____

5. In terms of the number of germ layers, how does a blastula differ from a gastrula?

UNIT XIV *Endocrine System*

PURPOSE   The purpose of Unit XIV is to enable the student to understand the physiology, microscopically recognize, and identify the location of the major endocrine glands.

OBJECTIVES   To complete Unit XIV, the student must be able to do the following:

1. Understand the basic physiology of selected endocrine glands and their hormones.
2. State the location of each endocrine gland in the body.
3. Using a microscope, recognize sections of pituitary, thyroid, parathyroid, pancreas, and adrenal glands.
4. List various pathological conditions associated with **hypersecretion** and **hyposecretion** of hormones by endocrine glands.

MATERIALS   microscope slides of the following endocrine glands: pituitary, thyroid, parathyroid, pancreas, and adrenal

PROCEDURE

Endocrine glands are ductless glands that secrete chemical agents known as **hormones**. Hormones from each endocrine gland are transported in the bloodstream to another part of the body where they evoke systemic responses or adjustments by acting on target tissues or organs. The endocrine glands together with the nervous system integrate functions of organs and systems in the body (see Figure 161 in Exercise 1).

The **pituitary gland** or **hypophysis** is located within the sella turcica of the sphenoid bone. It was at one time considered to be the master endocrine gland in the body. It does regulate the secretory functions of other endocrine glands, but it is now believed that the pituitary is regulated by the **hypothalamus** of the brain.

The anterior lobe of the pituitary is known as the **adenohypophysis** (or **pars distalis**). The adenohypophysis secretes several hormones, some of which are known as **tropic hormones** because they exert their effects indirectly by stimulating the functional activities of other endocrine glands. The major tropic hormones and the endocrine glands they affect (target glands) include the following:

1. **Thyroid stimulating hormone (TSH)** ⟶ thyroid gland.
2. **Adrenocorticotropic hormone (ACTH)** ⟶ adrenal cortex.
3. **Follicle stimulating hormone (FSH)** ⟶ ovaries and testes.
4. **Luteinizing hormone (LH)** ⟶ ovaries.
5. **Interstitial cell stimulating hormone (ICSH)** ⟶ testes (LH and ICSH are the same hormone).
6. **Luteotropic hormone (LTH)** ⟶ ovaries and mammary glands.

After a tropic hormone has been released by the pituitary gland, it stimulates its target endocrine gland to secrete its hormone(s). For example, TSH (thyroid stimulating hormone) stimulates the thyroid gland to produce the hormone thyroxine, which in the human regulates basal metabolism. The hypothalamus is thought to regulate the amount of TSH secreted by the adenohypophysis in response to the level of thyroxine in the bloodstream.

Other hormones secreted by the adenohypophysis are **somatotropin (STH)** or **growth hormone (GH)**, and **melanocyte stimulating hormone (MSH)**. STH is a general metabolic hormone having a variety of actions. It enhances growth of the cartilaginous portion of bone, causes an increase in muscle mass, decreases glomerular filtration rate, and liberates free fatty acids from adipose tissue. MSH exerts its action on melanocytes, the pigment producing cells in the skin. In the infant, MSH is found in the intermediate lobe of the pituitary gland. Later in life, the intermediate lobe becomes part of the neurohypophysis and continues MSH production.

The posterior lobe of the pituitary is known as the **neurohypophysis (or pars nervosa)**. The neurohypophysis secretes two hormones. The hormone **oxytocin** functions in stimulating uterine contractions and milk ejection. **Antidiuretic hormone (ADH)** promotes reabsorption of water from the distal convoluted tubules and collecting tubules in the kidney. ADH, when administered exogenously, is known as **vasopressin**. These two hormones are secreted by neurosecretory cells in the hypothalamus and transported by means of nerve fibers to the neurohypophysis where they are stored.

The **thyroid gland**, which secretes two hormones, is located anterior to the trachea and inferior to the larynx. Thyroid hormone, or **thyroxin**, regulates metabolic rate and is a derivative of **thyroglobulin**, a glycoprotein. The biosynthesis of thyroxin from thyroglobulin is relatively complex and involves several steps. **Thyrocalcitonin (or calcitonin)** is a hormone that lowers serum calcium levels and functions as an antagonist to parathyroid hormone.

The **parathyroid glands**, of which there are usually four, are located on the posterior surface of the lateral lobes of the thyroid. The parathyroids secrete **parathyroid hormone (PTH)** which increases serum calcium levels by mobilization of calcium from bone in response to lowered calcium levels in blood. Should the serum calcium level fall too low, a condition known as **tetany** develops, which may be fatal.

The **pancreas** has an endocrine function in addition to its **exocrine** function of secreting digestive enzymes, which you studied in a previous unit. Interspersed among the acinar regions of the pancreas are areas known as **islets of Langerhans**, which contain two cell types: **alpha**, which produce the hormone **glucagon**, and **beta**, which produce the hormone **insulin**. Insulin secretion regulates cellular uptake of glucose and blood glucose level. A lack of insulin results in **hyperglycemia** (elevated blood glucose level), mobilization of depot fat in adipose tissue into the blood, a decrease in protein synthesis, and dehydration. These symptoms characterize the disease **diabetes mellitus**. Glucagon raises the blood glucose level and is thought to promote glycogen breakdown to glucose in the liver. The major stimulant for the release of either insulin or glucagon is a change in the circulating blood glucose level. An increase in circulating blood glucose levels results in insulin release. Correspondingly, a decrease in blood glucose results in glucagon release.

There are two **adrenal glands (or suprarenal glands)**, one positioned on the superior border of each kidney. Each gland consists of a **cortex**, which secretes steroid hormones and surrounds the **medulla,** which secretes the **catecholamines epinephrine** and **norepinephrine.**

Steroids of the adrenal cortex include (1) **mineralocorticoids,** which increase sodium reabsorption by the distal convoluted tubules in the kidney; (2) **glucocorticoids,** which influence the metabolism of glucose, protein, and fat; and (3) **androgens,** which produce masculinization and are present in both sexes. The most abundant mineralocorticoid is **aldosterone;** glucocorticoids include **cortisol** and **corticosterone;** and the androgens, or male sex hormones, include the *17-keto steroids.* There is also evidence that **estrogens,** which are female hormones, are present in the adrenal cortex of both sexes.

EXERCISE 1   *Location of Endocrine Glands*

Label the endocrine glands in Figure 161.

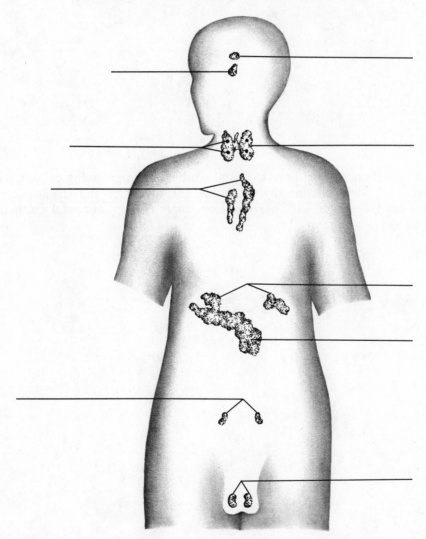

FIGURE 161 *The endocrine system*

EXERCISE 2   *Pituitary Gland (Hypophysis)*

Examine a section of pituitary gland. *Draw* the section and *label* the following: **infundibular stalk, pars distalis, chromophil cells, pars nervosa,** and **pars intermedia.** First observe this section under a dissecting microscope to see its entirety.

EXERCISE 3   *Thyroid Gland*

In this section, you will see cuboidal epithelium surrounding **colloid,** which stains pink and contains **thyroglobulin,** the precursor and storage form of thyroxin (see Figure 162). *Draw and label:* colloid, cuboidal epithelium.

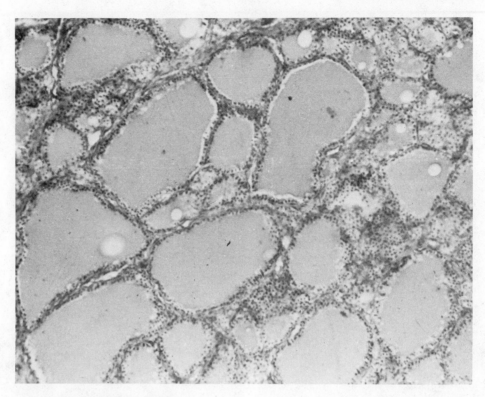

FIGURE 162 *Thyroid gland*

EXERCISE 4   *Parathyroid Glands*

Parathyroid tissue in the human adult is comprised of two principal types of cells: **chief** (or **principal**) cells and **oxyphilic** cells. Chief cells occur in masses whereas oxyphilic cells occur singly or in small groups. Oxyphils are not present in young children. *Draw* a section of parathyroid tissue, *labeling* oxyphilic cells, chief cells, and connective tissue.

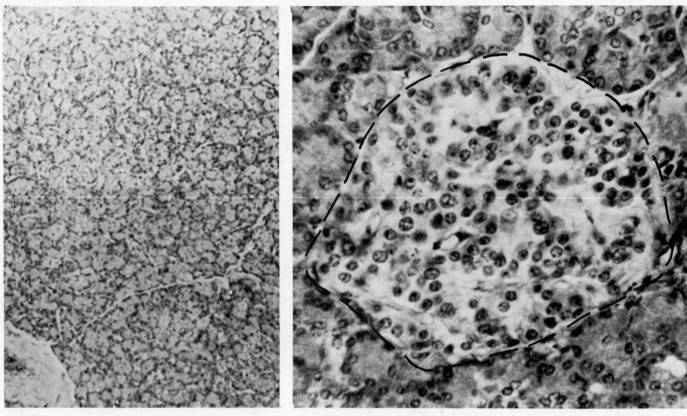

FIGURE 163   *General view of pancreas*   **FIGURE 164**   *Pancreas, islet of Langerhans*

EXERCISE 5   *The Pancreas*

Observe the scattered **islets of Langerhans** among the acinar tissue (Figures 163 and 164). Recall that the islets produce two hormones, glucagon from alpha cells and insulin from beta cells. *Draw* a section of pancreas, *labeling* the acinar and islet portions.

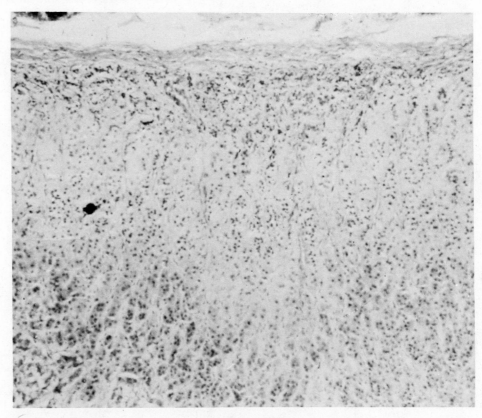

FIGURE 165 *View of the cortex and medulla of suprarenal gland*

EXERCISE 6　　*Adrenal Glands*

Each adrenal gland has an outer **cortex** surrounding the **medulla**. The cortex consists of three zones, the outer **zona glomerulosa**, which secretes aldosterone; the middle **zona fasciculata** and the inner **zona reticularis** which secrete glucocorticoids and sex hormones (see Figure 165).

Endocrine System

DISCUSSION

1. Name the hormone(s) secreted by each of these endocrine glands and identify this hormonal function:

    a. pituitary (anterior lobe) _____

    _____

    b. pituitary (posterior lobe) _____

    _____

    c. thyroid _____

    _____

    d. parathyroid _____

    _____

    e. pancreatic islets _____

    _____

    f. adrenal cortex _____

    _____

    g. adrenal medulla _____

    _____

2. Using your text or reference books, name the disease that may result from:

    a. removal of parathyroid glands _____

    b. hypersecretion of STH in an adult _____

    c. hyposecretion of insulin _____

    d. hypersecretion of corticosteroids _____

    e. hyposecretion of corticosteroids _____

    f. deficiency of thyroxin in an adult _____

    g. insufficient iodine in the diet _____

    h. hypothyroidism present from birth _____

i. hyposecretion of growth hormone in a child _____

j. hyposecretion of ADH _____

3. a. Does the human pituitary have a distinct pars intermedia? _____

_____

b. How does thyroglobulin differ from thyroxin functionally? _____

_____

c. If a thyroid gland were overactive, would you expect the follicular epithelium to retain its cuboidal shape?

_____ Explain.

GLOSSARY

A band *p. 87* Anisotropic band; myosin and actin filaments in a myofibril.

Abdominal ostium Anterior opening of fallopian tubes that serves as the port of entry for the ovum.

Abduct *p. 4* Movement away from body midline or from body part.

Acapnia Condition represented by a reduction of carbon dioxide in the blood.

Achilles tendon *p. 90* The tendon posterior to the heel, which connects the Gastrocnemius and Soleus muscles with the tuberosity of the calcaneus.

Acidophils (Eosinophils) *p. 139* Cells that are somewhat larger than neutrophils with an irregularly shaped and partially constricted bilobed nucleus. These cells constitute 2%-4% of the normal leukocyte population.

Acidosis Condition of blood represented by a reduction of blood bicarbonate/carbonic acid ratio.

Adduct *p. 4* Movement toward body midline or toward another body part.

Adenohypophysis *p. 259* See Pars distalis.

Adhesion *p. 4* Process by which parts come together on close approximation.

Adrenal glands *p. 228* Suprarenal glands, one located on superior border of each kidney.

Adrenalin *p.166* Epinephrine.

Adrenocorticotrophic hormone (ACTH) *p. 259* A hormone produced by the adenohypophysis, having its effect on the adrenal cortex.

Adventitia *p. 104* The outermost fibrous coat of the digestive tract.

Afferent *p. 186* Movement or conveyance toward a central point. A component leading toward a central point (e.g., afferent arteriole).

Agglutination (coagulation) *p. 159* The clumping together of erythrocytes.

Albustix *p. 233* A urine test for the presence of albumin.

Aldosterone *p. 261* A mineralocorticoid produced by the adrenal cortex.

Alkalosis Condition of blood represented by an increased blood bicarbonate/carbonic acid ratio.

Allantoic bladder *p. 228* An elongated, collapsed sac that lies between the umbilical arteries and serves as the fetal urinary bladder.

Alpha cells *p. 260* Cells in the islets of Langerhans that produce the hormone glucagon.

Alveolus *p. 178* Saclike structure found in lung and bone.

Amenorrhea Condition of absence or cessation of menses.

Amnion *p. 240* Thin, extraembryonic membrane surrounding each fetus in utero.

Anabolism Process by which simple substances are converted to complex compounds by living systems.

Analgesia Absence of sensitivity to pain.

Anaphase *p. 20* Mitotic phase in which homologous chromosomes move apart to opposite poles.

Anaphylaxis Hypersusceptibility to foreign protein.

Anastomosis A joining or union between separate structures—for instance, the joining of blood vessels.

Androgen *p. 261* A class of steroid hormones that produces masculinization.

Anesthesia *p. 197* Condition represented by loss of sensation.

Anoxia Condition of tissues represented by oxygen reduction below normal physiological levels.

Antibody (agglutinin) *p. 140* A protein produced in response to an invasion by a foreign substance.

Antidiuretic hormone *p. 260* See Vasopressin.

Antigen (agglutinogens) *p. 158* Any foreign substance that, when introduced into the body, will cause the production of an antibody that reacts specifically with the foreign substance.

Aperture A body opening or entrance.

Aphasia Condition represented by loss of speech and/or comprehension of speech, symbols.

Aponeurosis A sheath of connective tissue.

Apoplexy Condition represented by acute hemorrhage of the brain; this condition may also result from thrombosis and embolism.

Arbor vitae *p. 188* White matter of the cerebellum assuming a treelike shape.

Arthrosis Articulation of a joint.

Articulation Place or point of union, specifically between two or more bones.

Ascending colon *p. 99* Portion of large intestine between the cecum and transverse colon.

Ataxia Condition represented by defective muscular coordination.

ATP *p. 89* Adenosine triphosphate—a nucleotide contained in all cells; serves as energy reserve.

Atrium *p. 153* The upper chambers of the heart.

Atrophy A condition represented by a decrease in size and function of cells, tissues, organs, and organ parts.

Barfoed's test *p. 122* Test for presence of monosaccharides.

Basement membrane A supporting membrane underlying mucosal epithelium.

Basophils *p. 139* A granular leucocyte.

Benedict's test *p. 122* A test for presence of reducing sugars.

Beta cells *p. 260* Cells in the islets of Langerhans that secrete the hormone insulin.

Bicuspid *p. 96* A tooth posterior to the cuspid; having two points or cusps; premolar.

Bicuspid valve *p. 155* Valve consisting of two flaps or cusps of tissue between the left atrium and left ventricle; *mitral* valve.

Bile *p. 100* An alkaline fluid secreted by the liver and concentrated and stored in the gallbladder that facilitates the emulsification of fats.

Biuret reaction *p. 120* A reaction specific for compounds containing two peptide bonds united by a nitrogen or carbon atom.

Blastula *p. 247* A hollow ball of cells formed from a cleaving egg; a stage in embryological development.

Body The central portion of an organ.

Bolus A rounded mass; food that is swallowed or is in the superior part of the digestive tract.

Bowman's capsule *p. 230* First collection unit of a nephron, shaped as a hemisphere, with single-layered epithelial walls.

Brainstem *p. 192* Inferior section of the brain consisting of the midbrain, pons, and medulla oblongata.

Broad ligament *p. 240* One of the uterine ligaments.

Brunner's glands *p. 110* Mucous glands in the submucosa of the duodenum.

Buffer A fluid substance that tends to reduce changes in hydrogen ion concentration.

Bursa Saclike pouch that reduces friction between bones and tendons.

Calcitonin *p. 260* Thyrocalcitonin; hormone of the thyroid gland.

Calorie Unit of heat; amount of heat required to elevate 1 g of water 1°C.

Calyx *p. 230* Cup-shaped structure for urine collection in the renal medulla.

Canaliculus *p. 43* A small passageway; found in Haversian system of bone.

Cancellous Spongy; "soft" bone tissue.

Canine *p. 96* A conical pointed tooth, situated between the lateral incisor and first premolar; cuspid.

Carbohydrate *p. 121* An aldehyde or ketone derivative or a polyhydric alcohol; contains carbon, hydrogen, and oxygen.

Carboxyl group COOH—a radical group.

Carcinoma A malignant growth of cells.

Cardiac *p. 106* Pertaining to the heart; a muscle type occurring in the heart; portion of the stomach nearest the esophagus.

Cartilage *p. 30* A firm, flexible, supporting connective tissue.

Catecholamine *p. 261* A derivative of catechol; for example, epinephrine and norepinephrine.

Caudal *p. 1* Inferior in human anatomy, tending toward tail.

Cecum *p. 99* A blind pouch; first section of the large intestine.

Cell The structural and functional unit of the body.

Cementum *p. 100* A bony layer surrounding the dentin of the root of a tooth.

Centriole Structures within a centrosome that divide and function in spindle formation during cell division.

Centrosome A cellular organelle located outside the nucleus that contains centrioles.

Cerebellum *p. 189* Second largest brain division; concerned mainly with coordination of movements.

Cerebrum *p. 189* Largest brain division of man consisting of two hemispheres connected by the corpus callosum.

Cervix *p. 239* Neck or necklike structure; the inferior end of the uterus.

Chiasma *p. 189* X-shaped intersection as of the fibers of the optic nerve on the ventral brain surface.

Chief cells *p. 106* Cells in the cardiac and fundic region of stomach mucosa that secrete pepsinogen; cells in parathyroid tissue.

Chorion *p. 240* An extraembryonic membrane surrounding the fetal pig in utero.

Chordae tendineae *p. 151* Slender fibers attached to papillary muscles and cusps of the bicuspid and tricuspid valves. The papillary muscles and chordae tendineae serve to prevent the cusps of the valve from being forced open during systole.

Choroid *p. 214* Middle layer or tunic of the eye.

Chromatin *p. 20* Strands of DNA and protein in the nucleus of the cell that condense into chromosomes during cell division.

Chromophil cells *p. 262* Readily staining cells with granular cytoplasm in the adenohypophysis.

Chyme Semifluid material resulting from gastric digestion.

Cleavage *p. 247* Initial series of mitotic divisions of a fertilized egg.

Clinistix *p.233* Dipstick indicator for glucose used in urinalysis.

Clinitest *p. 233* A urinalysis test for the presence of glucose.

Clitoris *p. 239* A small erectile body, the size of a pea, located at anterior union of the vulva in the female; analogous to the penis in the male.

Cochlea *p. 219* A spiral-shaped portion of the inner ear.

Cohnheim's areas *p. 87* Areas within a muscle fiber that contain myofibrils, seen in cross section; Cohnheim's fields.

Collecting tubules *p. 231* Thick-walled straight tubules extending from nephron into the renal medulla.

Colloid *p. 262* A state of matter usually consisting of a liquid medium and a solute. The solute particles in contrast to crystalloids will not permeate a dialyzing membrane.

Colon *p. 99* The large intestine; bowel.

Colostrum Substance secreted by mammary glands as a function of pregnancy and parturition.

Columnar *p. 29* Tall, having greater length than width; epithelial type.

Columns *p. 230* Region between pyramids in the renal medulla.

Coma An abnormal state of reduced sensitivity.

Combistix *p. 233* Simple clinical method for indicating the presence of a number of substances not found in normal urine.

Common bile duct *p. 100* A duct formed by the cystic and hepatic ducts leading to the duodenum.

Condyle *p. 46* Bone marking—a rounded or knoblike projection.

Contraction period *p. 90* The time in which a muscle actually contracts or shortens.

Corium *p. 38* Dermis of the skin.

Cornea *p. 214* Anterior transparent area of the sclera of the eye.

Coronal *p. 1* A transverse plane dividing the body or any of its parts into anterior and posterior portions.

Corpus albicans *p. 243* A regressed corpus luteum.

Corpus cavernosum *p. 245* One of two lateral masses of erectile tissue found in the penis.

Corpus luteum *p. 243* A collapsed follicle that has expelled its ovum.

Corpus spongiosum *p. 245* A mass of erectile tissue in the penis through which the urethra passes (corpus cavernosum urethrae).

Cortex *p. 260* The outer portion of a structure.

Corticosterone *p. 261* A glucocorticoid; hormone of adrenal cortex.

Cortisol *p. 261* A glucocorticoid; hormone of adrenal cortex.

Costal *p. 4* Pertaining to the ribs.

Cranial *p. 1* Towards the head end of the body.

Crest *p. 46* Bone marking—an elevation or a ridge that serves as a point for muscle attachment.

Crown *p. 101* The exposed portion of a tooth.

Crystalloid A noncolloidal substance resembling a crystal that dissolves and forms a true solution; will disperse through a dialyzing membrane.

Cuboidal *p. 29* Cube shaped.

Cumulus oophorus *p. 243* A mound of cells immediately surrounding an ovum within a Graafian follicle.

Cuspid *p. 96* A conical pointed tooth, situated between the lateral incisor and first premolar; canine.

Cutaneous *p. 214* Having to do with the skin.

Cyanosis Blueness of the skin; indicates hypoxia.

Cystic duct *p. 100* A duct leading to the common bile duct connecting the gallbladder with the hepatic duct.

Cytoplasm *p. 20* The portion of a cell outside the nucleus and within the plasma membrane.

Deamination Removal of an amino group ($-NH_2$) from an amino acid.

Dehydration   Removal of water from the body or a tissue.

Dendrite *p. 185*   Short process extending from the nerve cell body to a presynaptic neuron.

Dense, fibrous *p. 29*   A type of connective tissue consisting mainly of parallel rows of bundles of fibers.

Dentin (Dentine) *p. 101*   A collagenous and calcified substance comprising the principal mass of a tooth.

Dermis *p. 38*   The layer of dense, vascular, connective tissue subjacent to the epidermis.

Descending colon *p. 99*   A portion of the large intestine lying in the vertical position on the left side of the abdomen, extending from the stomach to the level of the iliac crest.

Diabetes mellitus *p. 260*   A chronic metabolic disease resulting from insufficient insulin production by the pancreas.

Diabetes insipidus   A comparatively rare disease of pituitary origin that is characterized by the excretion of large volumes of urine.

Dialysis *p. 24*   The movement of crystalloids from colloids through a semipermeable membrane.

Diapedesis   The outward movement of formed elements of blood through intact walls of vessels.

Diaphragm *p. 97*   A skeletal muscle separating the thoracic cavity from the abdominal cavity; important in breathing.

Diarthrosis   A joint formation allowing free movement.

Diastole *p. 161*   The noncontractile phase of a heart chamber.

Diencephalon *p. 192*   Inferior mesial brain section consisting of the thalamus and hypothalamus; primarily of embryological significance.

Diffusion *p.23*   Movement of molecules from a highly concentrated area to a lower concentrated area until an equilibrium is achieved.

Digit   A finger or toe.

Dipeptide   Two amino acids linked together with a peptide bond.

Disaccharide *p. 122*   A sugar formed by the union of two monosaccharides.

Distal *p. 1*   Farthest from the trunk or point of origin of a part.

Distal convoluted tubule *p. 231*   Last tubular portion of a nephron; continuation of the loop of Henle.

Diurnal   Occurring daily.

Diverticulum   A pouch or pocket off of a tube or passage.

Dorsal *p. 1*   Pertaining to the back.

Double pith *p. 90*   Destruction of the brain and spinal cord for experimental purposes.

Ductus arteriosus (Botallo's Duct)   A duct that develops before birth as a shunt between the pulmonary trunk and aortic arch.

Ductus deferens *p. 241*   Vas deferens; sperm duct.

Ductus venosus   A fetal liver bypass that connects the left umbilical vein with the inferior vena cava.

Duodenum *p. 97*   The first portion of the small intestine, about 10 inches long.

Edema   Presence of extraordinary amounts of fluid in extracellular tissue spaces.

Effector   A nerve that activates muscular contraction or glandular secretion.

Efferent *p. 186*   Movement away from a central point. Also, component leading away from a central structure (e.g., efferent arteriole).

Electrocardiogram   A tracing of the electric current that initiates cardiac muscle contraction.

Electrolyte   A solution that is able to conduct electricity due to the presence of ions.

Embolus *p. 157*   A plug of undissolved matter that circulates in the blood stream, may be solid, liquid, or gaseous; frequently it may cause an obstruction within a blood vessel that is termed an *embolism*.

Embryo *p. 247*   An animal in early stages of development; the human being from implantation through the eighth week postconception.

Emulsified fat *p. 118*   Fat that has been converted into small fat droplets.

Enamel *p. 101*   A hard covering of the crown of a tooth, composed primarily of calcium salts.

Endocrine *p. 259*   Grandular organs that secrete hormones directly into blood.

Endometrium *p. 245*   The inner lining of the uterus.

Endoplasmic reticulum   A protein and lipid membraneous network in the cytoplasm through which biochemical components of the cell move.

Endothelium *p. 136*   Simple squamous epithelium that surrounds the lumen of blood vessels; innermost lining of blood vessels.

Enzyme *p. 124*   An organic catalyst.

Eosinophil *p. 139*   See Acidophil.

Epicondyle *p. 46*   Bone marking—a small rounded process above a condyle.

Epidermis *p. 38*   The outermost layer of skin consisting of five layers.

Epididymis *p. 241* A long coiled duct on the anterior lateral surface of each testis.

Epigastric *p. 4* An abdominal region lying superior to the stomach.

Epimysium *p. 87* Connective tissue surrounding a muscle.

Epinephrine *p. 261* A catecholamine secreted by the adrenal medulla; commercially known as Adrenalin.

Epiphysis *p. 47* Part of bone connected in embryological stage of life by cartilage to long bone; later becomes end portion of a long bone.

Epithelium *p. 29* A sheet of cells that covers an external surface or lines an internal surface, and functions in protection, absorption, secretion, and filtration.

Erythrocytes *p. 139* Red blood cells; have the shape of biconcave disks and have no nuclei; normal human males have 5.0 million per cubic millimeter of blood; females have approximately 4.0–5.0 million erythrocytes per cubic millimeter of blood.

Esophagus *p. 97* A muscular tube, about 10 inches long, extending from the pharynx to the stomach.

Estrogen *p. 261* A class of female hormones.

Evagination An out-pocketing of a part.

Exocrine *p. 260* A glandular organ that secretes outwardly through ducts.

External respiration Exchange of gases between the blood and external environment that occurs in the alveoli of the lungs.

Falciform ligament *p. 100* A ligament dividing the liver into right and left lobes, which are then further subdivided.

Fallopian tubes *p. 240* Paired tubes that serve as ducts from the ovaries to the uterus, through which ova pass.

Fascia *p. 75* Dense fibrous connective tissue covering muscles.

Fascicle, fasciculi (pl.) *p. 87* A subdivision of an entire muscle, a bundle of muscle fibers.

Fat *p. 123* See Lipid.

Fatty acid *p. 123* A saturated or unsaturated organic (monocarboxylic) acid.

Fertilization *p. 247* Union of sperm and egg; formation of a zygote.

Fertilization membrane *p. 247* A membrane found elevated from the egg surface at fertilization.

Fetus An unborn vertebrate; an unborn human being from the third month after conception until birth.

Fibrin *p. 157* An insoluble protein forming a mass of tangled strands in which red and white blood cells become enmeshed during coagulation.

Fibrous *p. 29* See Dense, fibrous.

First polar body *p. 241* A structure containing chromosomes formed by the meiotic division of a primary oocyte.

Fissure *p. 46* A bone marking—narrow slitlike opening.

Follicle *p. 241* A small excretory sac or gland.

Follicle stimulating hormone (FSH) *p. 259* A gonadotropic hormone whose target organ is the ovary or testis; produced by the anterior pituitary gland.

Follicular fluid *p. 241* Liquid within a Graafian follicle surrounding an ovum.

Foramen *p. 46* An opening through a bone for transmission of nerves and/or blood vessels.

Foramen oval An orifice that develops in the interatrial septum during fetal development. Failure to close after birth is known as atrial septal defect.

Fossa *p. 46* A bone marking—a cavity or hollow area in a bone.

Frontal *p. 3* A lengthwise plane dividing the body into anterior and posterior portions.

Fundus *p. 97* Enlarged portion of the stomach to the left of and superior to the cardioesophageal junction.

Gallbladder *p. 100* An accessory organ of the digestive tract that concentrates and stores bile, later ejecting it into the duodenum.

Ganglion *p. 166* A collection of nerve cells.

Gastric pits *p. 106* Narrow depressions containing glands extending through the full thickness of the gastric (stomach) mucosa.

Gastrohepatic ligament *p. 100* A band of connective tissue surrounding and supporting the common bile duct, attached to the anterior surface of stomach and posteromedial region of the hepatic lobes.

Gastrula *p. 247* An early developmental stage following the blastula stage, during which the germ layers are formed.

Glomerulus *p. 260* A cluster of blood vessels in close proximity to Bowman's capsule.

Glucagon *p. 260* A hormone produced by alpha-cells of the Islet of Langerhans which is involved in carbohydrate metabolism.

Glucocorticoids *p. 261* A class of adrenal cortical steroid hormones that influence the metabolism of glucose, protein, and fat.

Glyceride *p. 123* An ester (salt) of glycerol with fatty acids.

Glycerol *p. 123* An alcohol with the formula $C_3H_8O_3$ found in neutral fats.

Golgi apparatus An organelle comprised of membranes stacked upon each other, functioning in glycoprotein synthesis and the movement of substances to the external cell environment.

Gonadotropic hormone *p. 246* A type of hormone secreted by the adenohypophysis (pars distalis) that stimulates the gonads to secrete hormones.

Gonads *p. 239* Ovaries and testes; sex organs.

Graafian follicle *p. 241* A hollow fluid-containing sac surrounding a primary or secondary oocyte contained within an ovary.

Greater omentum *p. 97* A large fold of peritoneum covering the small and large intestines.

Gubernaculum *p. 241* A band of connective tissue connecting the testes to the scrotum in the male fetal pig.

Haustra Sacculation of wall of large intestine.

Head *p. 46* Rounded surface of bone connected to the shaft by the neck.

Hemastix *p. 233* A urine test for the presence of blood.

Hemopoietic *p. 40* Pertaining to blood cell formation.

Hemorrhage The escape of a large amount of blood.

Hemostasis *p. 140* An arrest or stoppage of escaping blood.

Hepatic duct *p. 100* A duct from the liver that joins the cystic duct to form the common bile duct.

Hepatic flexure *p. 99* The angle formed by the ascending and transverse colon.

Hilum *p. 228* Mesial indentation of kidney and lungs.

Homeostasis A tendency toward maintenance of stability in internal environment.

Horizontal *p. 3* A plane dividing the body or its parts into superior and inferior portions.

Hormone *p. 259* A chemical messenger secreted by an endocrine gland.

Humerus *p. 45* Longest and largest bone of upper arm.

Hyaline *p. 30* The most common type of cartilage; glassy in appearance.

Hydrolysis (see Formation of triglyceride) *p. 123* The splitting of a compound by the addition of water in which the hydroxyl group and hydrogen atom are incorporated in the different fragmented compounds.

Hyoid *p. 55* A bone that supports the base of tongue.

Hyperglycemia *p. 260* Elevated blood glucose level.

Hypersecretion *p. 259* Secretion of greater than normal amounts of a substance.

Hypertonic solution *p. 25* A solution that, in relation to another solution, contains a greater amount of dissolved material (solute).

Hypertrophy An enlargement or overgrowth of an organ or organ part.

Hypocalcemia The condition of having too low a blood calcium level.

Hypochondrium *p. 4* An abdominal region lying on either side of the epigastric regions and above the lumbar regions.

Hypogastric *p. 4* An abdominal region lying inferior to the umbilical region.

Hypophysis *p. 189* Pituitary gland.

Hyposecretion *p. 259* Secretion of less than normal amounts of a substance.

Hypothalamus *p. 259* A portion of the brain lying beneath the thalamus; contains autonomic regulatory centers.

Hypotonic solution *p. 25* A solution that, in relation to another solution, contains less dissolved material (solute).

Hypoxia A condition represented by a reduced oxygen content.

I band *p. 87* Isotropic band; actin filaments in a myofibril.

Ictotest *p. 233* Urine test for presence of bilirubin.

Ileum *p. 97* The third (last) and longest portion of the small intestine.

Iliac *p. 4* Abdominal regions lying on either side of the hypogastric region; relating to the ilium.

Incisor *p. 96* One of four front teeth in each jaw adapted for cutting.

Inferior *p. 1* Lower; away from the head.

Inferior nasal turbinates *p. 53* Paired facial bones in the nasal cavity.

Inflammation *p. 8* A physiological response of cells to injury or destruction by physical, chemical, or bacterial agents; symptoms of inflammation include swelling, redness, heat, and pain.

Infundibular stalk (infundibulum) *p. 189* A portion of the neurohypophysis that extends superiorly from the pars nervosa, connecting it with the hypothalamus.

Inguinal canal *p. 241* Paired canals in the groin through which spermatic cords pass prior to reaching the scrotum.

Insulin *p. 260* Hormone secreted by the beta cells of the islets of Langerhans that regulates cellular uptake of glucose and blood glucose level.

Intercalated disk *p. 88* Prominent markings at which point the ends of cardiac muscle fibers are joined together.

Internal respiration Metabolic reactions that involve oxygen consumption and carbon dioxide release.

Interphase *p. 20* The stage between successive mitotic divisions during which DNA replication takes place.

Interstitial cells of Leydig *p. 245* A group of cells between the seminiferous tubules that secrete testosterone.

Interstitial cell stimulating hormone (ICSH) *p. 259* A gonadotropic hormone whose target organ is the testis; LH in the female.

Intestinal glands *p. 109* Glands in the mucosa of the small and large intestines that secrete digestive enzymes and/or mucus; also called intestinal crypts of Lieberkühn.

Ion *p. 197* An atom that has a negative or positive charge, such as H+.

Iris *p. 214* Internal pigmented muscle of the eye that regulates the size of the pupil.

Irritability The ability of a tissue or an organism to respond to a stimulus from the environment, such as heat, cold, or pain.

Islets of Langerhans *p. 260* The endocrine portions of the pancreas.

Isotonic *p. 25* A solution that contains an equivalent amount of solute in relation to another solution.

Jejunum *p. 97* The second portion of the small intestine; between the duodenum and ileum.

Keratin *p. 38* A scleroprotein pigment that is a naturally occurring component of the skin, nails, and hair.

Ketone A chemical compound containing a carbonyl group; ketone bodies are compounds synthesized by the liver in the stepwise process of oxidation of fats.

Ketonuria The incomplete oxidation of fatty acids, giving rise to the accumulation of ketone bodies in the urine.

Ketostix *p. 233* A urine test for the presence of ketones, especially acetone.

Kidney *p. 227* Paired organs that function in the maintenance of normal blood constituents.

Kinesthetic (kinesthesia) The special sense of awareness of one's bodily positions and movements.

Kymograph *p. 200* Physiological recording apparatus consisting of a rotating drum and drive mechanism.

Lactase *p. 117* An enzyme that hydrolyzes lactose to glucose and galactose.

Lactose *p. 117* A disaccharide composed of glucose and galactose.

Lacuna *p. 43* An anatomical term referring to a small depression or hollow cavity, a depression containing osteocytes.

Lamella *p. 43* A thin bony plate, such as the Haversian lamella.

Lamina propria *p. 104* A connective tissue layer in the mucosa, underlying the mucosal epithelium and basement membrane.

Laryngopharynx *p. 96* A portion of the pharynx (throat) posterior to the larynx.

Latent period *p. 89* The shortest period of muscle contraction, occurs between the stimulus and contraction period.

Lateral *p. 1* Toward the side.

Lecithin A chemical compound of natural occurrence found in most animal tissues, especially egg yolk, semen, and nervous tissue; consists of glycerophosphoric acid and esters of stearic, oleic, or other fatty acids, combined with choline.

Lesion An inclusive term describing any damage to a particular group of tissues, such as ulcers, cataracts, eczema, or tuberculosis.

Leukemia *p. 143* A malignant disease of the spleen, bone marrow, and lymphoid tissue; characterized by wild proliferation of leukocytes with a subsequent reduction in the number of erythrocytes and blood platelets, which results in anemia and increased susceptibility to hemorrhage and infection.

Leukocytes *p. 139* White blood cells containing various shaped nuclei.

Ligament Dense fibrous bands of connective tissue that attach bones to bones.

Liminal stimulus *p. 90* Threshold stimulus; the amount of stimulus needed for a muscle to contract.

Line *p. 46* A bone marking—a slight ridge.

Lipase *p. 118* An enzyme that hydrolyzes emulsified fats to fatty acids and glycerol.

Lipid Fat; a class of compounds, including phospholipids, steroids, free fatty acids, and triglycerides.

Liver *p. 100* The largest gland in the body, located immediately inferior to the diaphragm, occupying most of the right hypochondriac region and part of the epigastric region.

Loop of Henle *p. 231* A portion of the nephron between the proximal and distal convoluted tubules that extends from the renal cortex to the medulla and then proceeds superiorly toward the renal cortex.

Lumbar vertebrae *p. 59* Five vertebrae inferior to the thoracic vertebrae in spinal column.

Luteinizing hormone (LH) *p. 259* A gonadotropic hormone whose target gland is the ovary in the female; called ICSH in male.

Luteotropic hormone (LTH) *p. 259* Tropic hormone secreted by the pars distalis and influences ovaries and mammary glands.

Lymphocytes *p. 139* A type of leukocyte that has a large, round, dark-staining nucleus surrounded by a thin rim of clear blue cytoplasm. Lymphocytes constitute 20%–25% of the normal leukocyte population.

Lysosome A membrane-bound organelle containing acid hydrolases; found in the cytoplasm.

Maltase *p. 117* An enzyme that hydrolyzes maltose to two glucose molecules.

Maltose *p. 117* A disaccharide comprised of two glucose units.

Matrix The basic background substance in which cells or tissues grow and develop, such as bone matrix.

Meatus *p. 46* A canal running within a bone.

Medial *p. 1* Toward the midline of the body.

Median sagittal *p. 3* A lengthwise plane from front to back dividing the body or any of its parts into equal right and left halves.

Medulla *p. 261* The inner portion of a structure.

Medulla oblongata *p. 189* Most inferior section of brain containing control centers for vital functions.

Medullary rays *p. 230* Structure within renal pyramids involved in urine release, leading toward papillae and calyces.

Megakaryocyte An enormous cell of bone marrow, having a lobulated nucleus, which gives rise to blood platelets.

Meiosis *p. 241* A process of nuclear division in which daughter cells are produced having half the number of chromosomes as the parent cell; this process involves two cell divisions, the second yielding the haploid number of chromosomes; specific to gametes.

Melanin *p. 38* A skin pigment formed in melanocytes in the stratum germinativum of the epidermis.

Melanocyte stimulating hormone (MSH) *p. 260* A hormone secreted by the pituitary gland that stimulates melanin-producing cells in the epidermis.

Meninges *p. 188* Covering of the brain and spinal cord.

Menopause The gradual waning of the menstrual cycle in the human female; during this period, the ovaries gradually cease to function resulting in the cessation of menstruation.

Menstruation *p. 245* The periodic sloughing off of the endometrial lining, accompanied by bleeding.

Mesentery *p. 97* A fold of parietal peritoneum that anchors abdominal organs to the dorsal body wall.

Mesial Toward the midline of the body.

Mesovarium *p. 240* Mesentery connecting the ovaries to the posterior aspect of the broad ligament.

Metabolism The total of all the chemical processes and reactions taking place in the living organism. Metabolism is composed of two phases: *anabolism,* the constructive phase, and *catabolism,* the destructive phase.

Metaphase *p. 20* A stage in mitosis during which the chromosomes are aligned across the equator of the cell.

Midbrain *p. 192* That portion of the brain containing the corpora quadrigemina and cerebral peduncles; also called the mesencephelon.

Millon reaction *p. 120* A reaction specific for the amino acid tyrosine.

Mineralocorticoids *p. 261* A class of steroid hormones secreted by the adrenal cortex that increases sodium reabsorption by the distal convoluted tubules in the kidney.

Mitochondrion A cytoplasmic membrane-bound organelle; contains oxidative enzymes.

Mitosis *p. 20* A process of cell division in which daughter cells are produced that have the same number of chromosomes as the parent cell.

Molar *p. 96* A tooth posterior to the bicuspids; adapted for grinding; tricuspid.

Molisch reaction *p. 122* A general test for carbohydrates.

Monocyte *p. 139* A granular leucocyte.

Monosaccharide *p. 122* A simple sugar.

Mucosa *p. 104* The innermost lining of the digestive tract.

Mucus The secretion of mucous membranes that contains water, mucin, and various inorganic salts.

Muscle fatigue *p. 91* The result of an accumulation of waste products such as lactic acid in a muscle.

Muscle fiber *p. 87* A muscle cell.

Muscle twitch *p. 89* A single muscle contraction.

Muscularis externa *p. 104* A layer of circular and longitudinal muscle beneath the submucosal layer of the digestive tract.

Muscularis mucosae *p. 104* A smooth muscle layer between the mucosa and submucosa.

Myeloid Relating to or derived from bone marrow.

Myofibril *p. 87* A group of myofilaments.

Myofilament *p. 87* A molecular threadlike contractile element within a myofibril; a contractile component of a myofibril.

Nasal *p. 54* Paired facial bones; form the bridge of nose.

Nasopharynx *p. 96* Portion of the throat located posterior and inferior to the nasal cavity.

Neck *p. 101* The portion of a tooth at the gum line where the crown and root meet; also constricted part of bone below head.

Necrosis *p. 140* The death of cells or tissues as a result of disease or injury.

Nephron *p. 231* The structural and functional unit of the kidney.

Neuroglia A supporting cell of the nervous system.

Neurohypophysis *p. 260* The posterior lobe of the pituitary gland.

Neuron *p. 185* Nerve cell responsible for generation, conduction, and transmission of a nervous impulse.

Neurula *p. 247* An early embryonic form marking the development of the neural tube and the appearance of the nervous system; embryological stage following the gastrula.

Neutrophils *p. 139* White blood cells that usually contain three to five irregularly oval-shaped lobes connected by thin strands of nuclear material. These cells constitute 65%-70% of the normal leukocyte population.

Ninhydrin reaction *p. 120* A test for alpha amino acids.

Nitrazine paper *p. 233* Chemically treated paper used as an indicator for pH.

Norepinephrine *p. 261* A hormone secreted by the adrenal medulla; also a sympathetic chemotransmitter.

Nuclear membrane A double protein-lipid membrane surrounding the nucleus.

Nucleolus *p. 19* A small dense structure in a nucleus consisting of RNA and protein.

Nucleus *p. 20* A structure containing DNA within a cell; control center of the cell.

Nystagmus *p. 206* Continuous involuntary movement of the eyeball.

Olfactory *p. 189* Relating to the sense of smell. The olfactory nerve—the first cranial nerve—is concerned with the sense of smell.

Oogenesis *p. 241* The process of egg formation by meiosis.

Oogonium *p. 241* An immature diploid ovum.

Ootid *p. 241* A haploid cell resulting from meiotic division.

Ophthalmic Relating to the eye.

Orifice An opening, entrance, or outlet of any body cavity or structure.

Oropharynx *p. 96* The portion of the throat located posterior to the mouth.

Osmosis *p. 24* The diffusion of water through a semipermeable membrane, from a region of greater concentration to a lesser concentration of solution.

Osseous *p. 30* A supporting connective tissue; bone tissue.

Ossification Pertaining to the process of formation or conversion into bone or osseous tissue.

Ovary *p. 240* A paired primary reproductive organ in the female that produces ova and female sex hormones.

Oviduct Fallopian tube.

Ovum *p. 241* An egg cell; a female gamete.

Oxidation The loss of electrons by atoms; the opposite reaction to reduction.

Oxyphilic cells *p. 263* Cells found in parathyroid tissue; more abundant in old age. Reserve cells; can produce parathormone.

Oxytocin *p. 260* A hormone stored by the neurohypophysis that stimulates uterine contractions and milk ejection.

Palatine *p. 54* Paired facial bones; form posterior one-third of roof of mouth.

Pancreas *p. 260* A tubuloacinar gland lying behind and below the stomach in the curve of the duodenum, having the exocrine function of secreting digestive enzymes and an endocrine function of producing the hormones insulin and glucagon.

Papillae *p. 230* Indentations in the renal medulla at the base of medullary rays.

Papillary muscles *p. 151* Cylindrical muscular bundles attached to the muscular wall of the ventricles.

Paralysis The loss or deterioration of movement of a limb or limbs of the body as a result of a wide variety of physical or emotional disorders.

Paranasal sinuses *p. 173* Air- or mucus-filled cavities surrounding the nose.

Parathyroid gland *p. 260* Several (usually four) endocrine glands located on the posterior surface of the lateral lobes of the thyroid gland.

Parathyroid hormone (parathormone) *p. 260* A hormone secreted by the parathyroid glands that increases serum calcium levels.

Parietal *p. 60* Paired cranial bones that form a portion of sides and roof of cranium.

Parotid *p. 96* A pair of salivary glands located anterior and inferior to the ear.

Pars distalis *p. 259* Anterior lobe of the pituitary gland.

Pars intermedia *p. 262* The intermediate lobe of the pituitary gland; virtually nonexistant in man.

Pars nervosa *p. 260* The neurohypophysis.

Parturition The act of expulsion of the fetus from the maternal uterus.

Peduncle A term designating a large group of nerve fibers traveling between different regions within the central nervous system.

Penis *p. 241* The male copulatory organ and urinary outlet.

Pepsin *p. 106* A protease in the stomach that hydrolyzes proteins to proteoses and peptones.

Pepsinogen *p. 106* The precursor of pepsin.

Peptidase *p. 117* An enzyme that hydrolyzes peptides to amino acids.

Peptone *p. 117* A partially digested protein.

Pericardium *p. 152* A protective sac of fibrous and serous tissue that encloses the heart.

Perimysium *p. 87* A connective tissue surrounding a fascicle.

Peyer's patches *p. 113* An aggregation of lymph nodules in the mucosa of the ileum.

pH The symbolic representation of hydrogen ion concentration, expressing the degree of acidity or alkalinity of a solution. A numerical scale from 0 to 14 is used to indicate the pH range.

Phagocyte *p. 140* A term applied to a cell that destroys microorganisms or foreign substances.

Pharynx *p. 96* The region posterior to nose, mouth, and larynx; throat.

Physiological apparatus A recording apparatus for muscle contraction, electrocardiograms, heart cycles, breathing rhythms, and other physiological functions.

Pithing *p. 90* The destruction of the brain (single pith) or brain and spinal cord (double pith) for experimental purposes.

Pitressin (vasopressin) Antidiuretic hormone; enhances water reabsorption from the distal convoluted and collecting tubules of the kidney.

Pituitary gland *p. 259* An endocrine gland located in the sella turcica of the sphenoid bone.

Placenta *p. 240* A structure attached to the uterine wall that enables the exchange of oxygen and other substances to take place between maternal and fetal blood.

Plasma *p. 139* The straw-colored fluid portion of the blood that constitutes 55% of the blood volume; transports inorganic and organic substances.

Plasma membrane *p. 20* The protein and lipid membrane surrounding a cell.

Platelet (thrombocyte) *p. 139* Small round-to-oval disks averaging 200,000–400,000 per cubic millimeter of blood; Platelets are associated with the clotting process.

Plexus *p. 195* A term applied to a network or group of veins or nerves.

Plicae circulares *p. 110* Elevations of the submucosa in the small intestine; the tallest plicae circulares are in the jejunum.

Polymorphonuclear *p. 139* Refers to a cell that possesses a deeply lobed nucleus that gives the appearance of being multilobed, (e.g.) a polymorphonuclear leukocyte.

Polypeptide *p. 117* A chain of amino acids connected by peptide bonds.

Pons *p. 189* Inferior brain section used as a bridge from medulla oblongata and spinal cord to higher brain divisions.

Premolar *p. 96* A tooth posterior to the canine; bicuspid.

Primary follicle *p. 241* A layer of cells surrounding an oogonium.

Primary oocyte *p. 241* The mature, diploid, potential ovum.

Primary spermatocytes *p. 243* The mature, diploid, potential sperm.

Principal cells *p. 263* Predominant cells of the parathyroid glands.

Process *p. 46* Bone marking—a general projection from a bone.

Prognosis A term applied to the probable outcome of a disease or disorder.

Prolapse A term applied to the inferior displacement of a part of the viscera.

Prophase *p. 20* A stage in mitosis during which chromosomes shorten and thicken, spindle fibers appear, and nucleolus and nuclear membrane disappear.

Prostate gland *p. 241* A gland inferior to the bladder in the male, which secretes an alkaline fluid that forms the greatest portion of seminal fluid.

Protease *p. 117* An enzyme that hydrolyzes proteins to proteoses, peptones, peptides, and amino acids.

Proteose *p. 117* A partially digested protein.

Protuberance *p. 46* A bone marking—a projection or eminence.

Proximal *p. 1* Nearest the trunk or point of origin of a part.

Proximal convoluted tubule *p. 231* The portion of a nephron between Bowman's capsule and loop of Henle.

Pseudostratified *p. 29* An epithelium with a layer of staggered cells, giving the appearance of several layers.

Ptosis *p. 204* Drooping of an organ, or part of an organ; especially a viscus or upper eyelid.

Pulp *p. 101* A soft vascularized core of tissue medial to the dentine portions of a tooth.

Pulse *p. 161* The alternating expansion and recoil of an artery.

Purkinje cells *p. 188* Large neurons at the border of the cerebellum and medulla.

Pus *p. 140* An opaque and viscid fluid characteristically produced in bacterial infections, which contains dead cell debris and tissue breakdown products.

Pyloric sphincter *p. 97* An annular muscle that prevents the backflow of food from the small intestine to the stomach.

Pylorus (pylorus) *p. 97* The inferior portion of the stomach, adjacent to the duodenum.

Pyramids *p. 230* The collections of medullary rays in the renal medulla of the kidney.

Ramus *p. 46* Bone marking—an armlike branch extending from body of bone.

Receptor *p. 214* The distal ends of dendrites of sense (sensory) neurons.

Rectum *p. 99* The last 7 inches of the large intestine.

Reflex *p. 197* An involuntary response to a stimulus.

Relaxation period *p. 90* The period of time following the muscle contraction period in which the muscle returns to its former length.

Renal capsule *p. 229* Fibrous connective tissue covering of the kidney.

Renal corpuscle *p. 230* A glomerulus and Bowman's capsule.

Renal cortex *p. 230* Periphery of the kidney containing nephrons and glomeruli.

Renal medulla *p. 230* The portion of the kidney subjacent to the cortex.

Renal pelvis *p. 230* Upper expanded end of the ureter.

Retina *p. 214* Innermost layer of the eye; contains rods and cones.

Rh factor A red blood cell antigen.

Ribosome The site of protein synthesis within a cell.

Root *p. 101* The portion of a tooth within the gum.

Rostrum A short cylindrical projection in the pig, fused with the upper lip, analogous to the snout, or nose.

Rouleaux formation *p. 139* The occurrence of a group of erythrocytes resembling a stack of coins within capillaries.

Rugae *p. 100* Longitudinal folds of the mucosa and submucosa in the stomach.

Sagittal *p. 3* A lengthwise plane from front to back, dividing the body or any of its parts into right and left sides.

Sarcomere *p. 87* The unit of histological structure and physiological action of a muscle fiber; extends from Z line to Z line.

Sciatic nerve *p. 90* A spinal nerve from the lumbosacral plexus; found in the posterior region of the leg.

Sclera *p. 214* The outermost layer of the eye.

Scrotum *p. 240* A skin-covered pouch in the male containing testes, epididymides, and the lower portion of spermatic cords; located in the perineal region.

Scrotum processus vaginalis *p. 240* Scrotum; protective muscular sac housing the testes and providing thermal regulation for testes.

Secondary oocyte *p. 241* A cell resulting from the meiotic division of a primary oocyte.

Secondary spermatocytes *p. 243* Cells formed by the meiotic division of primary spermatocytes.

Second polar body *p. 241* A structure containing a haploid number of chromosomes formed by the meiotic division of a secondary oocyte into a large ootid.

Seliwanow's test *p. 122* A test that differentiates between fructose and glucose.

Semilunar valves *p. 136* Valves that occur in the aorta and pulmonary artery; prevent the backflow of blood from the aorta and pulmonary artery into the left and right ventricles, respectively.

Seminiferous tubules *p. 245* Long coiled tubules within the testes in which sperm production and maturation take place.

Septum The thickened portion of the myocardium that separates the two atria and ventricles; the muscular wall dividing the atria is the interatrial septum and the wall dividing the ventricles is the ventricular septum; cartilaginous tissue separating the nostrils.

Serosa *p. 104* The outerlayer of an organ that is surfaced with a reflection of peritoneum.

Sigmoid colon *p. 99* Portion of the large intestine that extends from the level of the iliac crest to the rectum in the human being.

Simple *p. 30* One cell-layer thick.

Single pith *p. 90* Destruction of the brain for experimental purposes.

Sinus *p. 46* Bone marking—an irregularly shaped space often filled with air and lined with mucosal tissue.

Skeletal muscle *p. 40* Pertaining to the skeleton; striated voluntary muscle.

Smooth muscle *p. 40* An involuntary muscle type occurring primarily in the walls of viscera and blood vessels.

Solute A substance dissolved in a liquid.

Solution A liquid consisting of a solute and a solvent.

Solvent A liquid that dissolves another substance.

Somatotropin (STH) *p. 260* Growth hormone, secreted by the adenohypophysis.

Spasm An involuntary, abnormal muscular contraction.

Sperm *p. 245* A male gamete.

Spermatids *p. 242* Haploid cells formed by the division of secondary spermatocytes; when mature, they form sperm cells.

Spermatogenesis *p. 242* The process of sperm formation.

Spermatogonia *p. 242* Immature undifferentiated germ cells in the testis.

Spermatozoa *p. 242* Sperm.

Sphincter An annular muscle surrounding and capable of closing a body opening.

Sphygmomanometer *p. 161* An instrument used for the indirect measurement of arterial blood pressure.

Spine *p. 46* Bone marking—a sharp projection that serves as a point for muscle attachment.

Spiral colon Coiled section of colon anterior to ascending colon on left side of abdominal cavity.

Spleen *p. 100* An organ located in the left hypochondrium, the cells of which function in phagocytosis, hemopoiesis, and the storage of blood.

Splenic flexure *p. 99* The angle formed by the transverse and descending colon.

Squamous *p. 29* Flat; scalelike.

Staircase phenomenon *p. 90* The appearance of a recording of muscle contraction in which single stimuli of constant intensity are applied to a muscle, resulting in each twitch being slightly greater than the preceding one; Treppe or summation contraction.

Stimulus *p. 90* An agent that arouses or incites activity.

Stomach *p. 97* An elongated pouchlike structure that, in the human being, communicates superiorly with the esophagus and inferiorly with the duodenum.

Stratified *p. 30* Several cell layers thick.

Stratum corneum *p. 38* The most superficial layer of epidermis; it is keratinized.

Stratum germinativum *p. 38* The deepest layer of epidermis; contains melanocytes; cells undergo mitosis in this layer.

Stratum granulosum *p. 38* A thin epidermal layer beneath the stratum lucidum.

Stratum lucidum *p. 38* A clear translucent epidermal layer beneath the stratum corneum.

Stratum spinosum *p. 38* An epidermal layer beneath the stratum granulosum; contains "prickly" cells.

Subcutaneous Under the skin.

Subliminal *p. 90* A stimulus of less than threshold strength; subthreshold.

Sublingual *p. 96* Salivary glands located in the anterior midportion of the mouth, under the tongue.

Submandibular *p. 96* Under the mandible; submaxillary gland.

Submaxillary *p. 96* A pair of salivary glands located in the floor of the mouth, inferior to the mandible.

Submucosa *p. 104* The layer immediately underlying the mucosa in the digestive tract.

Subthreshold (subliminal) *p. 90* A stimulus of less than threshold strength.

Sucrase *p. 117* An enzyme that hydrolyzes sucrose to glucose and fructose.

Sucrose *p. 117* A disaccharide composed of glucose and fructose.

Superior *p. 1* Higher; toward the head end of the body.

Suprarenal gland *p. 260* Adrenal gland.

Synctium The fusion of cells that were originally separate.

Synovial fluid A lubricating fluid around a joint.

Systole *p. 161* The contraction phase of the cardiac cycle.

Telophase *p. 20* A stage in mitosis during which spindle fibers disappear, cytokinesis occurs, and the nucleus reforms.

Tendon *p. 40* A dense fibrous connective tissue band connecting bones to muscles.

Testis *p. 241* A paired, primary reproductive organ in the male, that produces sperm and testosterone.

Testosterone *p. 245* The primary male sex hormone.

Tetanus *p. 91* A condition in which a muscle is in a state of continual contraction.

Tetany *p. 91* A condition resulting from hypocalcemia.

Thoracic vertebrae *p. 59* Twelve vertebrae inferior to the cervical vertebrae in the vertebral column.

Thorax (thoracic) Chest region.

Threshold stimulus *p. 90* The amount of stimulus necessary to cause a response in a nerve, a muscle, and certain sensory receptors.

Thrombus *p. 140* A blood clot formed within a vessel.

Thyrocalcitonin *p. 260* A hormone secreted by the thyroid gland that lowers serum calcium levels.

Thyroglobulin *p. 260* A glycoprotein that is the storage form of thyroxin.

Thyroid gland *p. 260* An endocrine gland located in the neck, anterior to the trachea, inferior to the larynx.

Thyroid stimulating hormone (TSH) *p. 259* A tropic hormone produced by the adenohypophysis whose target gland is the thyroid.

Thyroxin *p. 260* The main thyroid hormone.

Tissue *p. 29* A group of cells performing a common function.

Transitional *p. 29* A type of stratified epithelium capable of stretching.

Transverse *p. 3* A plane dividing the body or its parts into superior and inferior portions.

Transverse colon *p. 99* The portion of the large intestine passing horizontally across the abdomen.

Treppe *p. 91* Staircase phenomenon in muscle contraction.

Tricuspid *p. 96* A tooth located posterior to the bicuspids having three cusps; molar.

Tricuspid valve *p. 151* A valve in the heart consisting of three flaps or cusps of tissue between the right atrium and right ventricle; prevents the backflow of blood from the ventricle into the atrium during systole.

Trochanter *p. 46* Bone marking—a large irregularly shaped projection to which muscle attaches.

Tropic hormone *p. 259* One of several hormones secreted by the adenohypophysis that stimulate other endocrine glands to secrete hormones.

Trypsin *p. 117* A protease in the pancreas that hydrolyzes proteins to proteoses, peptides, and amino acids under alkaline conditions.

Tubercle *p. 46* Bone marking—a small rounded process.

Tuberosity *p. 46* Bone marking—a large rough process, often smaller than a trochanter.

Tunica adventitia (externa) *p. 135* The outermost coat of blood vessels that contains a well-defined elastic layer with fibrous and areolar tissue supporting the nearby tissue.

Tunica intima *p. 135* The innermost coat of blood vessels composed of endothelium and fibroelastic tissue in longitudinal folds.

Tunica media *p. 135* The medial coat of blood vessels that contains smooth muscle and elastic fibers.

Umbilical *p. 4* The central region of the abdomen; pertaining to the umbilicus.

Ureter *p. 227* A 10- to 12-inch tube used for urine release connected superiorly to the kidneys and inferiorly to the bladder.

Urethra *p. 227* A urinary drainage duct from the bladder to the external orifice in the female, a joint urinary and genital canal in the male; a single tube leading to an external opening used for urine release; the terminal canal for urine release.

Urinary bladder *p. 227* A muscular reservoir for urine.

Urinary meatus (urethral orifice) The anterior-most opening of the female vestibule; used for urine release.

Urine *p. 232* Watery excretion containing nitrogenous wastes, urea, ammonia, pigments, and electrolytes.

Urogenital orifice *p. 239* Opening posterior to the genital papilla, through which pass products of the reproductive and urinary systems.

Urogenital sinus *p. 240* Cavity anterior to urogenital orifice in the female fetal pig that serves as canal for products of the reproductive and urinary systems.

Uterine tube Fallopian tube; oviduct.

Uterus *p. 241* A female secondary sex organ in which the embryo becomes implanted and develops prior to birth; womb, female receptacle for fetal growth and menstrual function.

Vagina *p. 239* A canal in the female that extends from the cervix to the external genital orifice.

Vasa vasorum *p. 135* Blood vessels supplying blood to the outer layers of the large blood vessels.

Vas deferens *p. 241* An extension of the epididymis through which sperm pass to the ejaculatory ducts; ductus deferens.

Vasopressin *p. 260* Antidiuretic hormone; Pitressin; secreted from the neurohypophysis.

Ventral *p. 1* Of or near the belly.

Ventricles *p. 151* The lower left and right chambers of the heart.

Vertebral column *p. 55* A collection of vertebrae through which the spinal cord passes.

Vertigo Dizziness.

Vestibule A body cavity that serves as an entrance to another cavity or space.

Villi *p. 110* Elevations of the mucosa of the small intestine.

Viscera (visceral) *p. 97* The internal organs of the abdomen.

Vulva The external female genitalia.

Z band *p. 87* A microscopic band that divides an I band in a myofibril; Z line.

Zona fasciculata *p. 265* The middle zone of the adrenal cortex; secretes glucocorticoids and sex hormones.

Zona glomerulosa *p. 265* The most peripheral portion of the adrenal cortex; secretes aldosterone.

Zona reticularis *p. 265* The inner portion of the adrenal cortex; secretes glucocorticoids and sex hormones.

SOLUTION APPENDIX

Some of the solutions below are available as commercially prepared solutions:

Barfoed's Solution

4.5 g crystallized neutral cupric acetate
100 ml distilled water
0.12 ml 50% acetic acid

Benedict's Solution

17.3 g $CuSO_4$
173.0 g sodium citrate
100.0 g Na_2CO_3 (anhydrous)
Dissolve citrate and carbonate by heating in 800 ml of distilled water.
Dissolve $CuSO_4$ in 100 ml of distilled water. Add $CuSO_4$ slowly to the first solution, stirring constantly. Add water to make a total of 1000 ml. May be kept indefinitely.

Holtfreter's Solution

Stock Solution:
3.5 g NaCl
0.05 g KCl
0.1 g $CaCl_2$
0.02 g $NaHCO_3$
Dissolve in 1 liter of H_2O (all glassware must be scrupulously clean).
When ready for use:
Add one (1) part stock solution to nine (9) parts distilled water.

Lugol's Iodine Solution

4.0 g iodine
6.0 g KI (Potassium iodide)
Dissolve in 100 ml of distilled water.

Methylene Blue Solution

10 mg methylene blue dissolved in 100 ml of distilled water.

Millon's Solution

Dissolve 100 g of Hg in 200 g of concentrated HNO_3.
Dilute solution with two volumes of water.

Molisch Reagent

5% alcoholic solution of alphanaphthol.

Ninhydrine Solution (Also commercially available as a spray)

0.1 g ninhydrine
100 ml distilled water.

Ringer's Solution (for frogs)

6.5 g NaCl
0.2 g $NaHCO_3$
0.1 g $CaCl_2$
0.1 g KCl
Add enough distilled water to make 1000 ml.

Saline Solution (Physiological)—humans

9.0 g NaCl dissolved in 1000 ml distilled water

Saline Solution (Physiological)—amphibians

7.0 g NaCl dissolved in 1000 ml distilled water.

Seliwanow's Reagent

0.5 g resorcinol
100 ml concentrated HCl
Add water to make a total of 300 ml.

ILLUSTRATION APPENDIX

FIGURE 6A Mitosis in onion root cells (cut A)
1. Late telophase

FIGURE 6C Mitosis in onion root cells (cut C)
2. Anaphase

FIGURE 12 Cuboidal epithelium as found in the collecting tubules of the kidney
1. Cuboidal cells

FIGURE 22 Hyaline cartilage of the trachea
1. Chondrocyte cells
2. Interterritorial matrix
3. Perichondrium

FIGURE 27 Thick, skin, palm, showing epidermal layers
3. Stratum granulosum

FIGURE 29 Skin, human scalp, showing hair follicles
1. Stratum corneum
2. Cortex of hair
3. Hair follicle
4. Hair bulb
5. Sebaceous gland

FIGURE 67 Tongue, showing taste buds
2. Taste bud

FIGURE 71 Stomach showing detail of mucosa, submucosa, and muscularis layers
1. Lumen of stomach
3. Arteriole
5. Submucosa layer

FIGURE 99 Sheep heart in situ, with pericardial sac removed, ventral aspect
1. Trachea
5. Ventricle, left
6. Apex

FIGURE 100 Sheep heart, in situ, with pericardial sac removed, dorsal aspect
1. Trachea
5. Left ventricle

FIGURE 101 Sheep heart with ventral portion reflected to show chamber detail
1. Trachea
7. Left ventricle
8. Apex

FIGURE 102 Sheep heart showing chamber and valve detail
2. Atrium, right
4. Ventricle, right

FIGURE 115 Sheep brain, superior aspect
3. Spinal cord
4. Longitudinal cerebral fissure

FIGURE 116 Sheep brain, inferior aspect, showing gross detail
1. Frontal lobes
5. Pons
7. Brainstem (spinal cord)
9. Temporal lobe

FIGURE 117 Sheep brain, lateral aspect, showing gross structures
4. Cerebellum
5. Medulla oblongata
6. Spinal cord

FIGURE 118 Sheep brain, midsaggital section, showing internal structures
6. Optic chiasma
7. Pituitary (hypophysis)
10. Medulla oblongata
11. Brainstem
15. Gray cortex of cerebellum

FIGURE 128 Sheep eye, midsaggital section, showing gross internal detail
3. Lens
4. Cornea
5. Pupil
8. Uvea (choroid)

FIGURE 129 Sheep eye, coronal section, showing internal detail, posterior to lens
1. Sclera
4. Uvea (choroid)

FIGURE 134 Sheep kidney, midsaggital section, showing internal detail
4. Ureter

3 4 5 6 7 8 9 10